Mathematical Statistical Mechanics

Mathematical Statistical Mechanics

COLIN J. THOMPSON

Princeton University Press
Princeton, New Jersey

To my parents, and to
Michal, Ben, and David

Published by Princeton University Press, Princeton, N.J.
In the U.K.: Princeton University Press, Guildford, Surrey

LCC 77-150071

ISBN 0-691-08219-7 (hardcover edition)
ISBN 0-691-08220-0 (paperback edition)

Originally published 1972 by The Macmillan Company

First Princeton University Press printing, 1979

Preface

This book is based on a one-semester course given to undergraduates in applied mathematics at M.I.T. and on a two-quarter course given to mathematics and physics graduate students at Northwestern. It is concerned with some mathematical aspects of kinetic theory, statistical mechanics, and biology. These subjects have, of course, been studied for many years, but only relatively recently by mathematicians. Consequently, most books on statistical mechanics, for example, emphasize the physical rather than the mathematical aspects of the subject. An exception is a recent book by Ruelle (B1969), *Statistical Mechanics: Rigorous Results*. Ruelle's book, however, is meant more for the expert than the student. It is hoped, therefore, that the present book will be a useful introductory textbook for mathematics students interested in applying their knowledge to some relatively new and exciting branches of physical mathematics, or for physics students with a mathematical bent. The contents by no means exhaust the entire subject of mathematical statistical mechanics. There are a number of omissions, notably quantum statistical mechanics, cluster expansions, and recent existence theorems and inequalities. The student interested in learning more about these and more advanced topics is referred to the books by Huang (B1963), Wannier (B1966), Uhlenbeck and Ford (B1963), and Ruelle (B1969).

In outline, the first chapter on kinetic theory begins with the fundamental work of Clausius, Maxwell, and Boltzmann and ends with the birth of statistical mechanics late last century. Thermodynamics, which is a necessary prerequisite for statistical mechanics, is the subject of Chapter 2. The presentation is brief and is included only for completeness. The Gibbs ensembles and the statistical mechanical basis for the laws of thermodynamics are presented in Chapter 3 along with recent work on the existence of the thermodynamic limit, or bulk thermodynamics. In Chapter 4, the classical theories of phase transitions and recent work on critical exponents are discussed.

The Ising model, which is probably the most studied model in theoretical physics, particularly in recent times as a model for phase transitions, is the subject of Chapters 5 and 6. The two-dimensional model in zero field has been solved exactly by a number of methods which fall into classes: algebraic methods, discussed in Chapter 5, and combinatorial methods, discussed in Chapter 6. The model has not been solved in three dimensions or in two dimensions with an external magnetic field. The algebraic and combinatorial questions posed by these and related unsolved problems are formidable, but hopefully they are questions that will be answered in the not too distant future. Apart from the mathematical interest in the model, there is considerable physical interest in it as a model for phase transitions. In the absence of any general satisfactory theory of phase transitions, it is hoped that by studying the Ising model, some insight into the intricacies of the critical region will emerge. The numerical analysis of the three-dimensional Ising model discussed in Chapter 6 has, in fact, partially fulfilled this hope. Chapter 7 concludes with some applications of the Ising model to biology. This forms part of the rapidly growing fields of biomathematics and biophysics, which are becoming more and more interesting and challenging fields to mathematicians and physicists.

In some instances it may be felt that the subject matter of the book would be better presented at the graduate level. With this in mind, some advanced material is included as appendixes. Although primarily aimed at students, it is hoped that this book may also be of value to mathematicians interested in learning statistical mechanics, and perhaps to physicists who are interested in learning the mathematical side of the subject. Consequently the book presupposes little knowledge of physics or mathematics, the only really necessary prerequisites being advanced calculus and an elementary knowledge of probability theory, linear algebra, and classical mechanics.

Finally, it is a pleasure to acknowledge the help and encouragement of various students and colleagues during the writing of this book. My point of view and understanding of statistical mechanics has come almost entirely from Professors M. Kac and G. E. Uhlenbeck, with whom I have had the good fortune to be associated over the past five or so years. The first three chapters in particular have been greatly influenced by their lectures, both published and unpublished, and I am particularly grateful to them for clarifying remarks on the zeroth law of thermodynamics. Dr. A. J. Guttmann kindly read the entire manuscript and made many helpful comments, suggestions, and corrections. I am also grateful to Professors R. B. Griffiths, D. Ruelle, and A. J. F. Siegert for their comments. It is a pleasure finally to thank Mrs. G. Goldberg, Mrs. J. Curtis, and Miss C. Davis for typing the manuscript.

C. J. T.

Contents

Note to the Reader

Reference books and articles are listed alphabetically in the Bibliography. In the text, Uhlenbeck and Ford (B1963), for example, refers to the book written by Uhlenbeck and Ford in 1963, and Kac (1956) refers to the article written by Kac in 1956.

CHAPTER 1

Kinetic Theory

1-1 Historical Sketch

Among the earliest recorded speculations on the nature of matter and the material world were those of Thales of Miletus (about 640–547 B.C.), many of whose ideas may have had their origin in ancient Egyptian times. He suggested that everything was composed of water and substances derived from water by physical transformation. About 500 B.C. Heraclitus suggested that the fundamental elements were earth, air, fire, and water and that none could be obtained from another by physical means. A little later, Democritus (460–370 B.C.) claimed that matter was composed of "atoms," or minute hard particles, moving as separate units in empty space and that there were as many kinds of particles as there were substances. Unfortunately, only second-hand accounts of these writings are available. Similar ideas were put forward by Epicurus (341–276 B.C.) but were disputed by Aristotle (384–322 B.C.), which no doubt accounts for the long delay in further discussions on kinetic theory.

Apart from a poem by Lucretius (A.D. 55), expounding the ideas of Epicurus, nothing further was done essentially until the seventeenth century, when

Gassandi examined some of the physical consequences of the atomic view of Democritus. He was able to explain a number of physical phenomena, including the three states of matter and the transition from one state to another; in a sense he fathered modern kinetic theory.

Several years later, Hooke (1635–1705) advanced similar ideas and suggested that the elasticity of a gas resulted from the impact of hard independent particles on the enclosure. He even attempted on this basis to explain Boyle's law (after Robert Boyle, 1627–1691), which states that if the volume of a vessel is allowed to change while the molecules and their energy of motion are kept the same, the pressure is inversely proportional to the volume. The first real contribution to modern kinetic theory, however, came from Daniel Bernoulli (1700–1782). Although he is often incorrectly credited with many of Gassandi's and Hooke's discoveries, he was the first to deduce Boyle's law from the hypothesis that gas pressure results from the impacts of the particles on the boundaries. He also foresaw the equivalence of heat and energy (about 100 years before anyone else realized it), so it is perhaps only just that Daniel Bernoulli has been given the title "father of kinetic theory." His work on kinetic theory is contained in his famous book *Hydrodynamica*, which was published in 1738. The interested reader can examine a short excerpt of this work in Volume 2 of *The World of Mathematics* [Newman (B1956)] or in Volume 1 of a three-volume work, *Kinetic Theory*, by S. G. Brush (B1965), where many other papers of historical interest can be found.

After Bernoulli there is little to record for almost a century. We then find Herapath (1821), Waterston (1845), Joule (1848), Krönig (1856), Clausius (1857), and Maxwell (1860) taking up the subject in rapid succession. From this point on we shall depart from a full history of development and concentrate on only the significant contributions from Clausius, Maxwell, and Boltzmann, who together paved the way for statistical mechanics.

1-2 The Krönig–Clausius Model

The Krönig–Clausius model [Krönig (1856) and Clausius (1857)] is perhaps the simplest possible model of an ideal gas and is described simply as follows.

Consider a cube of side l containing N molecules each of mass m. We denote by $V = l^3$ the volume of the cube and by $A = l^2$ the area of each of the six faces (see Figure 1.1), and we make the following

ASSUMPTION. *The molecules are evenly divided into six uniform beams moving with velocity c in each of the six coordinate directions.*

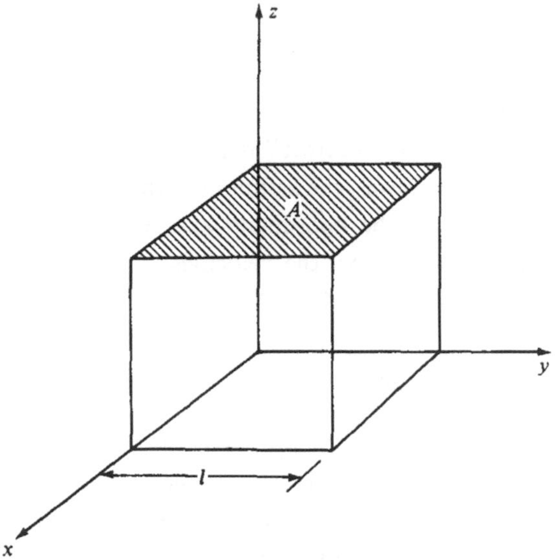

FIGURE 1.1. Cube of side l. Molecules are evenly divided into six uniform beams moving with velocity c in each of the six coordinate directions.

The problem is to find the pressure the molecules exert on each of the six faces of the cube. Since pressure is force per unit area and force is momentum per unit time, we have

$$\text{pressure} = \frac{\text{momentum}}{\text{unit area} \times \text{unit time}}. \tag{2.1}$$

In time δt, only the fraction

$$\frac{c\,\delta t}{l} N \tag{2.2}$$

of the molecules can strike a face. Only one sixth of these are moving toward any given face, and each molecule imparts momentum $2mc$ to the face it strikes, so the pressure P exerted on a face is

$$\begin{aligned} P &= \frac{1}{6}\frac{c\,\delta t}{l} N \frac{2mc}{l^2\,\delta t} \\ &= \frac{2}{3}\left(\frac{1}{2}Nmc^2\right)V^{-1}. \end{aligned} \tag{2.3}$$

If we now identify $\frac{1}{2}Nmc^2$ with the total kinetic energy E we obtain *Boyle's law*:

$$PV = \tfrac{2}{3}E. \tag{2.4}$$

We stress that from the mathematical point of view this model is irreproachable. Moreover, it yields the "correct" relation between P and V. From the physical point of view, however, the assumption that all molecules move with uniform speed parallel to the coordinate axes is clearly absurd. A more realistic assumption would allow the velocities to be described by a probability density. This leads us to Maxwell's model.

1-3 The Maxwell Distribution

In 1859 Maxwell [Maxwell (1860)] proposed the following model: *Assume* that the velocity components of N molecules, enclosed in a cube with side l, along each of the three coordinate axes are independently and identically distributed according to the density $f(\alpha) = f(-\alpha)$, i.e.,

$$f(v_i)\,dv_i = \text{the probability that the } i\text{th velocity component is between } v_i \text{ and } v_i + dv_i, \tag{3.1}$$

where i = 1, 2, 3 refer to the three coordinate axes. As before, only the fraction

$$\frac{v_i\,\delta t}{l}N \tag{3.2}$$

of the molecules whose ith velocity component is v_i can strike a face perpendicular to the i axis in time δt. Hence the pressure exerted on such a face is

$$\begin{aligned} P &= \int_0^\infty f(v_i)\,\frac{v_i\,\delta t N}{l}\,\frac{2mv_i}{l^2\,\delta t}\,dv_i \\ &= \frac{2N}{V}\int_0^\infty mv_i^2 f(v_i)\,dv_i \qquad (3.3) \\ &= \frac{2N}{V}\langle \tfrac{1}{2}mv_i^2\rangle, \end{aligned}$$

where $\langle\cdots\rangle$ denotes the probabilistic average with respect to the distribution $f(\alpha)$, i.e.,

$$\langle g(v_i)\rangle = \int_{-\infty}^{\infty} g(v_i)f(v_i)\,dv_i. \tag{3.4}$$

Now, since we are assuming that each component has the same distribution,

$$\left\langle \frac{1}{2} m v_i^2 \right\rangle = \frac{1}{3} \left\langle \frac{1}{2} m \sum_{i=1}^{3} v_i^2 \right\rangle$$

$$= \frac{1}{3N} \left\langle \frac{1}{2} N m v^2 \right\rangle \tag{3.5}$$

$$= \frac{1}{3N} \langle E \rangle,$$

where $\langle E \rangle$ is the average total kinetic energy of the molecules and $v^2 = v_1^2 + v_2^2 + v_3^2$. Thus Equation 3.3 gives

$$PV = \tfrac{2}{3}\langle E \rangle, \tag{3.6}$$

which again is Boyle's law.

Maxwell actually went one step further and assumed in addition to Equation 3.1 that the distribution depends only on the magnitude of the velocity; in other words,

$$f(v_1)f(v_2)f(v_3)\, dv_1\, dv_2\, dv_3 = \phi(v_1^2 + v_2^2 + v_3^2)\, dv_1\, dv_2\, dv_3, \tag{3.7}$$

from which it follows immediately that

$$f(v_i) = A \exp(-Bv_i^2)$$

$$\text{or} \qquad \phi(v_1^2 + v_2^2 + v_3^2) = A^3 \exp[-B(v_1^2 + v_2^2 + v_3^2)], \tag{3.8}$$

which is the famous Maxwell distribution for velocities.

Again, from the mathematical point of view, the derivation of Equations 3.6 and 3.8 is perfectly legitimate, granted the assumptions. However, even though the assumptions represent a vast improvement on Clausius' assumption, they still may be objected to on physical grounds. The main objection is that from the dynamical point of view, the velocities do not enter independently into the equations describing collisions between molecules, so it is reasonable to expect that the velocities of different molecules are correlated in some way. Maxwell himself realized this and briefly returned to the problem in his great memoir entitled "On the Dynamical Theory of Gases" [Maxwell (1867)], in which he proposed: "It is the velocities of two colliding molecules, rather than the velocity components of a single molecule, that are statistically independent." By considering two colliding molecules and using the fact that the total kinetic energy is conserved in a collision, Maxwell was again able to show that his distribution followed from this perhaps more realistic assumption.

At this time Ludwig Boltzmann (1844–1906) was just beginning his career in theoretical physics. Inspired perhaps by Maxwell's memoir, he became immersed in Maxwell's assumptions and the role collisions play in bringing about equilibrium. This work finally culminated in his great work of 1872 [Boltzmann (1872)], which contained the now-famous "Boltzmann equation." A derivation of this equation is given in the following section.

1-4 The Boltzmann Equation

Unlike Maxwell, who presupposed equilibrium and looked for analytical conditions on the distribution function required to maintain stable equilibrium, Boltzmann started out by assuming that the gas was not in equilibrium and attempted to show that equilibrium can result from collisions between molecules. To this end he derived his famous equation.

For simplicity, we consider, as Boltzmann did, a spacially homogeneous gas of N hard-sphere particles (i.e., perfectly elastic billiard balls), enclosed in a volume V, with mass m, diameter a, and velocity distribution function $f(\mathbf{v}, t)$ defined by

$$f(\mathbf{v}, t)\, d^3v = \text{number of particles at time } t \text{ with velocity in the volume element } d^3v \text{ around } \mathbf{v}. \tag{4.1}$$

We make the following assumptions:

1. Only binary collisions occur. That is, situations in which three or more particles come together simultaneously are excluded. Physically speaking, this is a reasonable assumption if the gas is sufficiently dilute (i.e., if the mean free path is much larger than the average molecular size).

2. The distribution function for pairs of molecules is given by

$$f^{(2)}(\mathbf{v}_1, \mathbf{v}_2, t) = f(\mathbf{v}_1, t) f(\mathbf{v}_2, t). \tag{4.2}$$

In other words, the number of pairs of molecules at time t, $f^{(2)}(\mathbf{v}_1, \mathbf{v}_2, t)\, d^3v_1\, d^3v_2$ with velocities in d^3v_1, around $\mathbf{v}_1$ and in d^3v_2 around $\mathbf{v}_2$, respectively, is equal to the product of $f(\mathbf{v}_1, t)\, d^3v_1$ and $f(\mathbf{v}_2, t)\, d^3v_2$. This is Boltzmann's famous *Stosszahlansatz*, or assumption of "molecular chaos," and it is this innocent-looking assumption [originally due to Clausius (1857)] that was, and still is some hundred or so years later, the most widely discussed.

It is obvious that assumption 1 is purely a dynamical assumption and that 2 is basically a statistical assumption. Unfortunately, the latter was not clearly or adequately stressed, particularly by Boltzmann at a time when objections to his equation were based solely on mechanical considerations. We shall discuss this point at length in later sections after first deriving here Boltzmann's equation for $f(\mathbf{v}, t)$.

By definition, the rate of change of $f(\mathbf{v}, t)\, d^3v$ is equal to the net gain of molecules in d^3v as a result of collisions, i.e.,

$$\frac{\partial f}{\partial t} = n_{\text{in}} - n_{\text{out}}, \tag{4.3}$$

where

$$n_{\text{in(out)}}\, d^3v = \text{the number of binary collisions at time } t \text{ in which one of the final (initial) molecules is in } d^3v. \tag{4.4}$$

Consider now a particular molecule [1] with velocity in d^3v_1, around $\mathbf{v}_1$, and all those molecules [2] with velocities in d^3v_2 around $\mathbf{v}_2$ which can collide with molecule [1]. It is to be remembered that we are considering now only binary collisions of hard spheres with diameter a.

In the relative coordinate system with molecule [1] at rest, the center of molecule [2] must be in the "collision cylinder," as shown in Figure 1.2, if it is to collide with molecule [1] during the time interval δt. The collision cylinder and appropriate parameters are shown in more detail in Figure 1.3. The collision cylinder has volume $b\, d\phi\, db\, g\delta t$, and from Figure 1.2 it is obvious that

$$b = a \cos\left(\frac{\theta}{2}\right), \tag{4.5}$$

and hence that

$$\begin{aligned} b\, db\, d\phi &= a \cos\left(\frac{\theta}{2}\right) \left|\frac{db}{d\theta}\right| d\theta\, d\phi \\ &= \frac{a^2}{2} \cos\left(\frac{\theta}{2}\right) \sin\left(\frac{\theta}{2}\right) d\theta\, d\phi \qquad (4.6) \\ &= \frac{a^2}{4}\, d\Omega, \end{aligned}$$

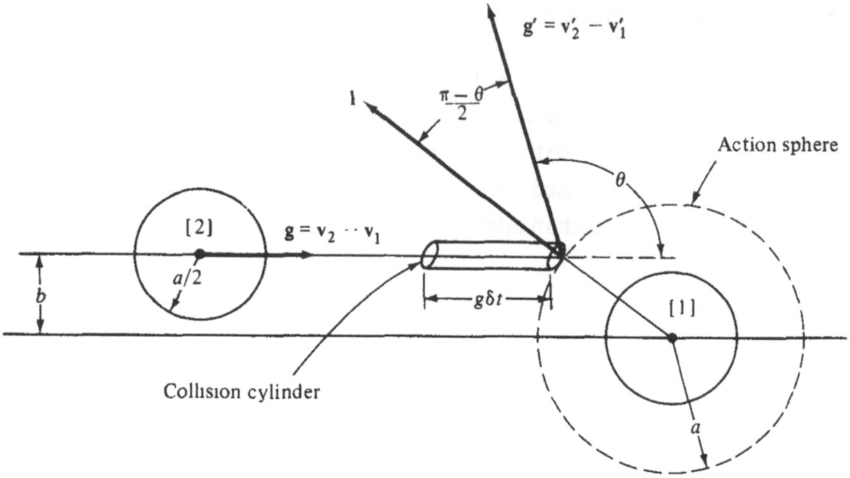

Figure 1.2. Collision of two hard spheres with initial velocities $\mathbf{v}_1$ and $\mathbf{v}_2$ and final velocities $\mathbf{v}_1'$ and $\mathbf{v}_2'$ respectively, shown in the relative coordinate system with sphere [1] at rest. $\mathbf{l}$ is the unit vector in the direction of the line of centers at the time of collision.

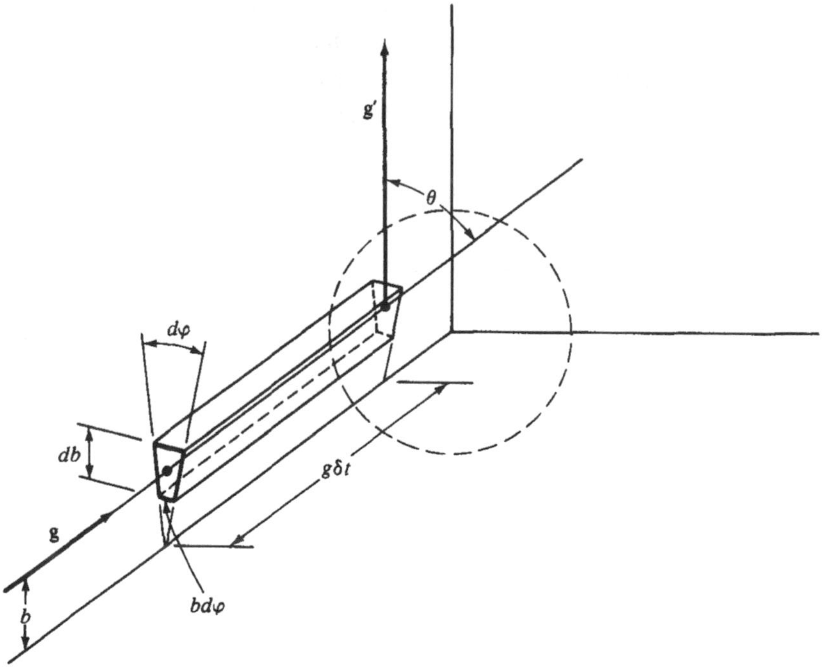

Figure 1.3. Collision cylinder, action sphere, and appropriate parameters for two colliding hard spheres.

where $d\Omega = \sin\theta \, d\theta \, d\phi$ is the solid angle element of the "scattered" particle [2].

Now using the definition of n_{out} (Equation 4.4), and the Stosszahlansatz Equation (4.2) we have

$$n_{out}\, d^3v_1 = \frac{a^2}{4} \int d\Omega \int |\mathbf{v}_2 - \mathbf{v}_1| [f(\mathbf{v}_2, t)\, d^3v_2][f(\mathbf{v}_1, t)\, d^3v_1]. \tag{4.7}$$

To determine n_{in} we merely look at the inverse collision $(\mathbf{v}_1', \mathbf{v}_2') \to (\mathbf{v}_1, \mathbf{v}_2)$ and use the above results to obtain immediately

$$n_{in}\, d^3v_1 = \frac{a^2}{4} \int d\Omega \int |\mathbf{v}_2' - \mathbf{v}_1'| [f(\mathbf{v}_2', t)\, d^3v_2'][f(\mathbf{v}_1', t)\, d^3v_1']. \tag{4.8}$$

It is to be noted that the second integral in Equation 4.7 is over $\mathbf{v}_2$ and in Equation 4.8 over $\mathbf{v}_2'$.

Since we are considering perfectly elastic spheres, energy ($\frac{1}{2}mv^2$) and momentum ($m\mathbf{v}$) are conserved in a collision; hence since all masses are assumed to be equal,

$$\mathbf{v}_1 + \mathbf{v}_2 = \mathbf{v}_1' + \mathbf{v}_2' \qquad \text{(momentum conservation)} \tag{4.9}$$

and

$$v_1^2 + v_2^2 = v_1'^2 + v_2'^2 \qquad \text{(energy conservation).} \tag{4.10}$$

It follows almost immediately (see Problem 3) that

$$|\mathbf{v}_2' - \mathbf{v}_1'| = |\mathbf{v}_2 - \mathbf{v}_1| \tag{4.11}$$

and that the transformation $(\mathbf{v}_1, \mathbf{v}_2) \to (\mathbf{v}_1', \mathbf{v}_2')$ is orthogonal, i.e.,

$$d^3v_1'\, d^3v_2' = d^3v_1\, d^3v_2 . \tag{4.12}$$

In fact, it is not difficult (see Problem 3) to derive the following explicit relations from Equations 4.9 and 4.10:

$$\mathbf{v}_1' = \mathbf{v}_1 + [(\mathbf{v}_2 - \mathbf{v}_1) \cdot \mathbf{l}]\mathbf{l} \tag{4.13}$$

and

$$\mathbf{v}_2' = \mathbf{v}_2 - [(\mathbf{v}_2 - \mathbf{v}_1) \cdot \mathbf{l}]\mathbf{l}, \tag{4.14}$$

where, as in Figure 1.2, $\mathbf{l}$ is the unit vector in the direction of the line of centers of the two spheres at the time of collision.

Using the results 4.11 and 4.12 combined with Equations 4.7 and 4.8 for n_{in} and n_{out}, the equation of motion (4.3) becomes

$$\frac{\partial f_1}{\partial t} = \frac{a^2}{4} \int d\Omega \int d^3v_2 |\mathbf{v}_2 - \mathbf{v}_1| (f_1' f_2' - f_1 f_2), \tag{4.15}$$

where $f_1 = f(\mathbf{v}_1, t)$, $f_2' = f(\mathbf{v}_2', t)$, etc. This is *Boltzmann's equation*, which can clearly be generalized in a straightforward way: first, for more general interactions (than hard spheres) and, second, for nonuniform systems.

For more general interactions, we write (see Equation 4.6)

$$b\, db\, d\phi = I(g, \theta)\, d\Omega \qquad (g = |\mathbf{v}_2 - \mathbf{v}_1|), \tag{4.16}$$

which simply defines the "differential scattering cross section" $I(g, \theta)$. For hard spheres $I(g, \theta) = a^2/4$, so for more general interactions, $(a^2/4)\, d\Omega$ on the right-hand side of Equation 4.15 is replaced by $I(g, \theta)\, d\Omega$.

For nonuniform systems, we consider $f(\mathbf{r}, \mathbf{v}, t)$ defined by

$f(\mathbf{r}, \mathbf{v}, t)\, d^3r\, d^3v =$ the number of molecules which at time t have positions in the volume element d^3r around $\mathbf{r}$, and velocities in the velocity–space element d^3v around $\mathbf{v}$. (4.17)

In the presence of an external force $\mathbf{F}(\mathbf{r})$ the point $(\mathbf{r}, \mathbf{v}, t)$ becomes

$$\left(\mathbf{r} + \mathbf{v}\, \delta t, \mathbf{v} + \frac{\mathbf{F}}{m} \delta t, t + \delta t\right) \text{ after time } \delta t.$$

The left-hand side of Equations 4.3 and 4.15 is then replaced by

$$\lim_{\delta t \to 0} \frac{\left[f\left(\mathbf{r} + \mathbf{v}_1\, \delta t, \mathbf{v}_1 + \frac{\mathbf{F}}{m} \delta t, t + \delta t\right) - f(\mathbf{r}, \mathbf{v}_1, t) \right]}{\delta t}$$
$$= \frac{\partial f_1}{\partial t} + \mathbf{v}_1 \cdot \nabla_{\mathbf{r}} f_1 + \frac{\mathbf{F}}{m} \cdot \nabla_{\mathbf{v}} f_1 \tag{4.18}$$

where $\nabla_{\mathbf{r}}$ denotes gradient with respect to $\mathbf{r}$ and $\nabla_{\mathbf{v}}$ gradient with respect to $\mathbf{v}$. The right-hand side of Equation 4.15 stays the same [with either $(a^2/4)\, d\Omega$ or $I(g, \theta)\, d\Omega$] with f_1 defined by $f_1 = (f(\mathbf{r}, \mathbf{v}_1, t)$, etc.

It is to be noted that in the nonuniform case the Stosszahlansatz states that in a given volume element d^3r the number of pairs of molecules with respective velocities lying in d^3v_1 around $\mathbf{v}_1$ and d^3v_2 around $\mathbf{v}_2$ is given by

$$[f(\mathbf{r}, \mathbf{v}_1, t)\, d^3r\, d^3v_1][f(\mathbf{r}, \mathbf{v}_2, t)\, d^3r\, d^3v_2]. \tag{4.19}$$

In a loose sense, one could say that this implies that the velocity of a molecule is uncorrelated with its position. Or, more precisely, the velocities of pairs of molecules are statistically independent.

It is obvious that Boltzmann's equation (4.15) or its generalizations discussed above, is hopelessly difficult to solve in general. By a very beautiful trick, however, Boltzmann was able to show that Equation 4.15 leads to equilibrium and that the equilibrium distribution is the Maxwell distribution. This is the subject of the next section. We shall not discuss here methods for solving Boltzmann's equation. This has become almost a science in itself and leads to a number of interesting mathematical questions. Since the main aim of this chapter is to provide background for equilibrium statistical mechanics, we shall concern ourselves only, in later sections of this chapter, with the validity of Boltzmann's equation and the role of the Stosszahlansatz in bringing about equilibrium. The interested reader is referred to Chapman and Cowling (B1953) and Cercignani (B1969) for excellent accounts of mathematical problems arising directly from the Boltzmann equation.

1-5 Boltzmann's *H* Theorem, Entropy, and Information

The equilibrium distribution is defined to be the time-independent solution of Boltzmann's equation (4.15). For simplicity, we consider the spacially uniform case so that $f_{\text{equilibrium}} = f_0(\mathbf{v})$. The left-hand side of Equation 4.15 then vanishes identically and Boltzmann's equation becomes, for general interactions,

$$\int d\Omega \int d^3v_2\, gI(g, \theta)[f_0(\mathbf{v}_2')f_0(\mathbf{v}_1') - f_0(\mathbf{v}_2)f_0(\mathbf{v}_1)] = 0, \tag{5.1}$$

where $g = |\mathbf{v}_2 - \mathbf{v}_1|$ and $I(g, \theta)$ is defined by Equation 4.16. For Equation 5.1 to be satisfied, it is clearly sufficient that

$$f_0(\mathbf{v}_2')f_0(\mathbf{v}_1') = f_0(\mathbf{v}_2)f_0(\mathbf{v}_1). \tag{5.2}$$

It was a great triumph for Boltzmann that he was able to show that Equation 5.2 is also a necessary condition for equilibrium. This is a consequence of

Boltzmann's *H* Theorem. *If $f(\mathbf{v}, t)$ is a solution of Equation* 4.15 *and $H(t)$ is defined by*

$$H(t) = \int d^3v f(\mathbf{v}, t) \log f(\mathbf{v}, t), \tag{5.3}$$

then

$$\frac{dH}{dt} \leq 0. \tag{5.4}$$

Proof. Note first of all from Equation 5.3 that

$$\frac{dH}{dt} = \int d^3v \frac{\partial f}{\partial t}(1 + \log f). \tag{5.5}$$

Clearly $\partial f/\partial t = 0$ implies that $dH/dt = 0$; hence $dH/dt = 0$ is a necessary condition for equilibrium.

Substituting Equation 4.15 into Equation 5.5 gives

$$\frac{dH}{dt} = \int d^3v_1 \int d^3v_2 \int d\Omega\, gI(g, \theta)(f_2' f_1' - f_2 f_1)(1 + \log f_1). \tag{5.6}$$

Since $gI(g, \theta)$ is invariant under interchange of $\mathbf{v}_1$ and $\mathbf{v}_2$, Equation 5.6 is unchanged by interchanging $\mathbf{v}_1$ and $\mathbf{v}_2$. Thus

$$\frac{dH}{dt} = \int d^3v_2 \int d^3v_1 \int d\Omega\, gI(g, \theta)(f_2' f_1' - f_2 f_1)(1 + \log f_2). \tag{5.7}$$

Adding Equations 5.6 and 5.7 and dividing by 2 then gives

$$\frac{dH}{dt} = \frac{1}{2}\int d^3v_1 \int d^3v_2 \int d\Omega\, gI(g, \theta)(f_2' f_1' - f_2 f_1)(2 + \log f_1 f_2). \tag{5.8}$$

We now change variables $(\mathbf{v}_1, \mathbf{v}_2) \to (\mathbf{v}_1', \mathbf{v}_2')$. We have seen in the previous section [Equations (4.11) and (4.12)] that $d^3v_1'\, d^3v_2' = d^3v_1\, d^3v_2$ and that $g = |\mathbf{v}_2 - \mathbf{v}_1| = |\mathbf{v}_2' - \mathbf{v}_1'|$; hence

$$\frac{dH}{dt} = \frac{1}{2}\int d^3v_1 \int d^3v_2 \int d\Omega\, gI(g, \theta)(f_2 f_1 - f_2' f_1')(2 + \log f_1' f_2'). \tag{5.9}$$

Taking one half of the sum of Equations 5.8 and 5.9 then gives

$$\frac{dH}{dt} = \frac{1}{4}\int d^3v_1 \int d^3v_2 \int d\Omega\, gI(g, \theta)(f_2' f_1' - f_2 f_1)(\log f_1 f_2 - \log f_1' f_2'). \tag{5.10}$$

Now it is easily shown that for any positive real numbers x and y,

$$(y - x)(\log x - \log y) \leq 0, \tag{5.11}$$

with equality holding if and only if $x = y$. The proof of the theorem is then complete when we note from Equation 5.11 that the integrand in Equation 5.10 is never positive.

We deduce immediately from Equations 5.10 and 5.11 that $dH/dt = 0$ if and only if the integrand of Equation 5.10 vanishes identically, i.e.,

$$f_1'f_2' = f_1 f_2, \tag{5.12}$$

which is precisely Equation 5.2. The proof also shows, since $H(t)$ is bounded below (see Problem 4) and monotonic nonincreasing, that under arbitrary initial conditions $\lim_{t\to\infty} f(\mathbf{v}, t) = f_0(\mathbf{v})$ defined by Equations 5.2 and 5.12. Equilibrium may, of course, be reached in a finite time.

The observant reader will have noticed that the above "proof" requires the existence and uniqueness of a solution of Boltzmann's equation, and in addition requires that all integrals converge. Both conditions need not, and probably are not, satisfied in general. The only case of which the author is aware, in which complete rigor has been supplied, is hard spheres. This was done by Carleman (B1957). The generalization of Carleman's analysis to systems with finite-range forces is probably not difficult, and further generalization would not go unnoticed!

Returning to Equation 5.2, it is now an elementary matter to deduce that the equilibrium distribution is the Maxwell–Boltzmann distribution. Before doing so, let us digress for a moment and examine the connection between Boltzmann's H function, entropy, and information. The relation is as follows: When the system is in equilibrium, $-H$ is proportional to the entropy, so if one is willing to define entropy as being proportional to $-H$ for nonequilibrium states, then the H theorem can be interpreted as a generalized second law of thermodynamics (see Section 2-4); i.e., entropy increases unless the system is in equilibrium. This also suggests a connection between entropy and probability. In particular, entropy is a measure of disorder or randomness, the equilibrium state being the most random distribution. Stated another way, entropy is a measure of our own information about the positions and velocities of the particles, so if we know only the total number of particles and the total energy, the equilibrium distribution corresponds to specifying the minimum amount of information [for more information see Jaynes (1957)].

1-6 The Maxwell–Boltzmann Distribution

We have seen from Boltzmann's H theorem that the equilibrium state, in the absence of external forces, is specified by the distribution $f_0(\mathbf{v})$ satisfying

$$f_0(\mathbf{v}_1)f_0(\mathbf{v}_2) = f_0(\mathbf{v}_1')f_0(\mathbf{v}_2'), \tag{6.1}$$

where $(\mathbf{v}_1, \mathbf{v}_2)$ and $(\mathbf{v}_1', \mathbf{v}_2')$ are the initial and final velocities, respectively, of any possible binary collision.

Taking logarithms of both sides of Equation 6.1 we have

$$\log f_0(\mathbf{v}_1) + \log f_0(\mathbf{v}_2) = \log f_0(\mathbf{v}_1') + \log f_0(\mathbf{v}_2'), \tag{6.2}$$

which has the form of a *conservation law*. Since for spinless molecules (e.g., hard spheres) the only conserved quantities are energy and momentum (and constants), it follows that $\log f_0(\mathbf{v})$ must be a linear combination of v^2 and the three components of $\mathbf{v}$, plus an arbitrary constant, i.e.,

$$\log f_0(\mathbf{v}) = \log A - B(\mathbf{v} - \mathbf{v}_0)^2,$$

or

$$f_0(\mathbf{v}) = A \exp[-B(\mathbf{v} - \mathbf{v}_0)^2], \tag{6.3}$$

where A, B, and the three components of $\mathbf{v}_0$ are arbitrary constants. Equation 6.3 is Maxwell's distribution for velocities.

We now express the arbitrary constants in Equation 6.3 in terms of physical quantities.

Since by definition the total number of particles N is given by

$$N = \int d^3r \int d^3v f(\mathbf{r}, \mathbf{v}, t), \tag{6.4}$$

we have in equilibrium that

$$\begin{aligned} \frac{N}{V} &= \int d^3v\, f_0(\mathbf{v}) \\ &= A \int d^3v \exp[-B(\mathbf{v} - \mathbf{v}_0)^2] \\ &= A \left(\frac{\pi}{B}\right)^{3/2}, \end{aligned} \tag{6.5}$$

where V is the volume of the container. It follows that $B > 0$ and

$$A = \left(\frac{B}{\pi}\right)^{3/2} \left(\frac{N}{V}\right). \tag{6.6}$$

The average velocity $\langle \mathbf{v} \rangle$ of a gas molecule is given by

$$\begin{aligned}\langle \mathbf{v} \rangle &= \frac{\int \mathbf{v} f_0(\mathbf{v})\, d^3v}{\int f_0(\mathbf{v})\, d^3v} \\ &= \frac{AV}{N} \int d^3v \mathbf{v} \exp[-B(\mathbf{v} - \mathbf{v}_0)^2] \\ &= \frac{AV}{N} \int d^3v (\mathbf{v} + \mathbf{v}_0) \exp(-B\mathbf{v}^2) \\ &= \mathbf{v}_0 .\end{aligned} \tag{6.7}$$

Thus if the gas has no translational motion ("Galilean invariance") $\mathbf{v}_0 = \mathbf{0}$.

The average kinetic energy ε of a molecule when $\mathbf{v}_0 = \mathbf{0}$ is given by

$$\begin{aligned}\varepsilon &= \frac{\int d^3v \frac{1}{2} m\mathbf{v}^2 f_0(\mathbf{v})}{\int d^3v f_0(\mathbf{v})} \\ &= \frac{mAV}{2N} \int v^2 \exp(-Bv^2)\, d^3v \\ &= \frac{3m}{4B}\end{aligned} \tag{6.8}$$

i.e.,

$$B = \frac{3m}{4\varepsilon}, \tag{6.9}$$

and, from Equation 6.6,

$$A = \left(\frac{3m}{4\pi\varepsilon}\right)^{3/2} \frac{N}{V}. \tag{6.10}$$

Now the equation of state (see Equation 3.6) is

$$PV = \tfrac{2}{3} N\varepsilon \tag{6.11}$$

and experimentally temperature is defined by $PV = NkT$, where k is Boltzmann's constant and T the absolute temperature, i.e., $\varepsilon = \frac{3}{2}kT$, and, from Equations 6.10, 6.9, and 6.3,

$$f_0(\mathbf{v}) = \frac{N}{V} \left(\frac{m}{2\pi kT}\right)^{3/2} \exp\left(-\frac{m\mathbf{v}^2}{2kT}\right). \tag{6.12}$$

In deriving Equation 6.12 we have neglected external forces, so let us now consider the equilibrium distribution for a dilute gas in the presence of an external conservative force field **F** given by

$$\mathbf{F} = -\nabla\phi(\mathbf{r}). \tag{6.13}$$

In this case the equilibrium distribution function is

$$f_0(\mathbf{r}, \mathbf{v}) = f_0(\mathbf{v})\exp\left[-\frac{\phi(\mathbf{r})}{kT}\right] \tag{6.14}$$

($\mathbf{v}_0$ need not be zero now). To prove this, we note that Equation 6.14 satisfies Boltzmann's equation (4.15) with the left-hand side replaced by Equation 4.18. Since $f_0(\mathbf{v})$ satisfies Equation 5.2, the right-hand side of Equation 4.15 is zero, and it is a trivial matter to show (see Problem 5) from Equations 6.12, 6.13, and 6.14 that the left-hand side of Equation 4.15,

$$\left(\mathbf{v}\cdot\nabla_{\mathbf{r}} + \frac{\mathbf{F}}{m}\cdot\nabla_{\mathbf{v}}\right) f(\mathbf{r}, \mathbf{v}) = 0, \tag{6.15}$$

which completes the proof.

The complete Maxwell–Boltzmann distribution (6.14) was first written down by Boltzmann (1868), who derived the equation from some generalizations of Maxwell's arguments. The derivation given here is also due to Boltzmann (1872).

1-7 Time Reversal, Poincaré Cycles, and the Paradoxes of Loschmidt and Zermelo

Shortly after Boltzmann published his H theorem, it was pointed out by Kelvin and a little later by Loschmidt (1876) that if molecular collisions are governed by Newtonian mechanics, any given sequence of collisions can be run backward as well as forward. Boltzmann's H theorem, however, singles out a preferred direction in time. In addition to this "reversal paradox," there is also the "recurrence paradox," which follows from a theorem of Poincaré:

POINCARÉ'S THEOREM: *A mechanical system enclosed in a finite volume and with a finite energy will, after a finite time, return to an arbitarily small neighborhood of almost any given initial state.* (The precise meaning of "neighborhood" and "almost" will be stated in a moment.)

Zermelo (1896) pointed out that Poincaré's theorem implies that Boltzmann's H function is a quasi-periodic function of time and hence that a deterministic mechanical system cannot remain in a final state, as one might have expected from the H theorem.

Since the proof of Poincaré's theorem is so simple and elegant, we shall give it here before discussing the paradoxes in some detail in the next section.

We first need a few preliminaries. A *state* of a gas is determined by the $3N$ canonical coordinates $q_1, q_2, \ldots, q_{3N}$ and the $3N$ conjugate momenta $p_1, p_2, \ldots, p_{3N}$ of the N molecules (assuming each molecule has only three degrees of freedom). The $6N$-dimensional space spanned by these vectors $(p, q) = (p_1, p_2, \ldots, p_{3N}; q_1, q_2, \ldots, q_{3N})$ is called the Γ space, or phase space, of the system. A point in Γ space then represents a state of the entire system.

Now let

$$\rho(p, q, t)\, d^{3N}p\, d^{3N}q = \text{the number of points which at time } t \text{ are contained in the } \Gamma\text{-space volume element } d^{3N}p\, d^{3N}q \text{ around the point } (p, q). \tag{7.1}$$

Given $\rho(p, q, t)$ at any time t, its future (and past) values are determined by Hamilton's equations of motion, which are

$$\left.\begin{aligned} \dot{p}_i &= \frac{dp_i}{dt} = -\frac{\partial H}{\partial q_i} \\ \dot{q}_i &= \frac{dq_i}{dt} = \frac{\partial H}{\partial p_i} \end{aligned}\right\} \quad i = 1, 2, \ldots, 3N, \tag{7.2}$$

where $H = H(p, q)$ is the Hamiltonian of the system (i.e., kinetic energy plus potential energy). We shall consider throughout only Hamiltonians of the form

$$H(p, q) = \sum_{i=1}^{3N} \frac{p_i^2}{2m} + V(q_1, q_2, \ldots, q_{3N}). \tag{7.3}$$

Various restrictions on the potential-energy function V will be imposed later.

Hamilton's equations (7.2) tell us how points move in Γ space as time evolves. For Hamiltonians that do not depend on any time derivatives of p and q, such as Equation 7.3, Equations 7.2 are invariant under time reversal. It is clear also that if the coordinates of a point in Γ space are given at any time, Equations 7.2 determine the motion of the point *uniquely* for all other times.

For systems with constant finite energy E, the motion of a point in Γ space is restricted to the *energy surface* $H(p, q) = E$. We now define the measure

$\mu(A)$ of a region A on the energy surface to be the normalized area of A, i.e. (see Appendix A),

$$\mu(A) = \frac{\int_A d\sigma \|\text{grad } H\|^{-1}}{\int_\Omega d\sigma \|\text{grad } H\|^{-1}} \tag{7.4}$$

where $d\sigma$ denotes an element of energy surface, Ω denotes the entire energy surface $[\mu(\Omega) = 1]$, and

$$\|\text{grad } H\|^2 = \sum_{i=1}^{3N} \left[\left(\frac{\partial H}{\partial p_i}\right)^2 + \left(\frac{\partial H}{\partial q_i}\right)^2\right]. \tag{7.5}$$

Poincaré's theorem may now be stated precisely as follows: For an arbitrary (small) set A denote by B the subset of A consisting of those points which never return to A once they have left A. The measure of the set $\mu(B) = 0$, which is to say that almost all points in A return to A after a finite time. Restricting the system to a finite volume with finite energy is essential in the proof and simply means that the measure of the entire energy surface is finite.

Before we proceed with the proof we need one lemma, which is the celebrated

LIOUVILLE THEOREM. *If $\rho(p, q, t)$ is defined by Equation 7.1, then*

$$\frac{d\rho}{dt} = \frac{\partial \rho}{\partial t} + \sum_{i=1}^{3N} \left(\frac{\partial \rho}{\partial p_i}\frac{dp_i}{dt} + \frac{\partial \rho}{\partial q_i}\frac{dq_i}{dt}\right) = 0. \tag{7.6}$$

This means that if we picture points moving in Γ space as a fluid, the flow is incompressible. Also, and of most use to us, we have as a consequence that the "volume" of a set of points flowing in Γ space is independent of time. We have given a detailed proof of this measure-theoretic statement of Liouville's theorem in Appendix A. Heuristically, it is almost obvious. Thus, if we consider a small region V of Γ space which is small enough so that $\rho(p, q, t)$ is uniform (i.e., independent of p and q in V), the number of points ΔN in V is independent of time if V evolves according to Hamilton's equations; hence

$$\frac{d}{dt}(\Delta N) = \frac{d}{dt}\rho \int_V d\Gamma = \frac{d\rho}{dt}\int_V d\Gamma + \rho\frac{d}{dt}\int_V d\Gamma = 0, \tag{7.7}$$

where $d\Gamma = dp_1 \cdots dp_{3N}\, dq_1 \cdots dq_{3N}$. But in view of Equation 7.6, the first term on the right-hand side of Equation 7.7 vanishes; hence

$$\frac{d}{dt}\int_V d\Gamma = 0. \tag{7.8}$$

Since small volumes V in Equation 7.8 can be pieced together to form big volumes, Equation 7.8 holds for arbitrary V. Now by considering two adjacent energy surfaces E and $E + dE$ and letting $dE \to 0$ we deduce from Equation 7.8 that if $\mu(A)$ is defined by Equation 7.4 and A_t denotes the time evolution of region A, then

$$\mu(A_t) = \mu(A). \tag{7.9}$$

This form of Liouville's theorem is exactly what we need to prove Poincaré's theorem. Let us pause for a moment, however, to prove statement 7.6.

Consider an arbitrary volume element ω in Γ space and let S denote its surface. Since the number of points leaving ω per unit time is equal to the rate of decrease of the points in ω,

$$\frac{d}{dt}\int_\omega \rho\, d\Gamma = \int_S \mathbf{n} \cdot \mathbf{v}\rho\, dS, \tag{7.10}$$

where $\mathbf{n}$ is the unit vector normal to S and $\mathbf{v}$ is the $6N$-dimensional "velocity":

$$\mathbf{v} = (\dot{p}_1, \dot{p}_2, \ldots, \dot{p}_{3N}, \dot{q}_1, \dot{q}_2, \ldots, \dot{q}_{3N}). \tag{7.11}$$

From Equation 7.10 and the divergence theorem (in $6N$-dimensions) we immediately deduce the continuity equation,

$$\frac{\partial \rho}{\partial t} + \nabla \cdot (\rho \mathbf{v}) = 0, \tag{7.12}$$

where ∇ is the $6N$-dimensional gradient operator

$$\nabla = \left(\frac{\partial}{\partial p_1}, \frac{\partial}{\partial p_2}, \ldots, \frac{\partial}{\partial p_{3N}}, \frac{\partial}{\partial q_1}, \frac{\partial}{\partial q_2}, \ldots, \frac{\partial}{\partial q_{3N}}\right). \tag{7.13}$$

From Equation 7.12 it follows that

$$\begin{aligned} -\frac{\partial \rho}{\partial t} &= \nabla \cdot (\rho \mathbf{v}) \\ &= \sum_{i=1}^{3N} \left[\frac{\partial}{\partial p_i}(\rho \dot{p}_i) + \frac{\partial}{\partial q_i}(\rho \dot{q}_i)\right] \qquad (7.14) \\ &= \sum_{i=1}^{3N} \left[\frac{\partial \rho}{\partial p_i}\dot{p}_i + \frac{\partial \rho}{\partial q_i}\dot{q}_i\right] + \sum_{i=1}^{3N} \rho\left[\frac{\partial}{\partial p_i}(\dot{p}_i) + \frac{\partial}{\partial q_i}(\dot{q}_i)\right], \end{aligned}$$

and since in view of Equations 7.2 the last term in Equation 7.14 vanishes, the result (7.6) follows.

Finally, we prove Poincaré's theorem. Let A be a region of Γ space and B the subset of A containing points which never return to A. We assume that $\mu(B)$, defined by Equation 7.4, is finite and establish a contradiction.

Consider the sequence of regions $B_1, B_2, B_3, \ldots$ which evolve from B after times t, $2t$, $3t$, ..., respectively. Clearly we can choose t large enough so that regions B and B_1 do not overlap, i.e., $\mu(B \cap B_1) = 0$. From this it follows that all the regions B_i are nonoverlapping, for suppose that B_n and B_{n+k} have points in common: Since paths of points in Γ space cannot intersect, it follows that B_{n-1} and B_{n+k-1} have points in common and hence that B and B_k have points in common. This means that there are points in B which return after a time kt. Since this contradicts our initial assumption about B, $B, B_1, B_2, \ldots$ are nonoverlapping regions. Now from Liouville's theorem (Equation 7.9)

$$\mu(B) = \mu(B_1) = \mu(B_2) = \cdots, \tag{7.15}$$

so if $\mu(B)$ is finite, the measure of the whole energy surface must be unbounded, which establishes the required contradiction. $\mu(B)$ is therefore zero and the proof of Poincaré's theorem is complete.

The time it actually takes, by the way, to return to an initial state is called a *Poincaré cycle.*

1-8 Validity of the Boltzmann Equation

The derivation of the Boltzmann equation (4.15) was based on two assumptions, one *mechanical* (binary collisions) and the other *statistical* (the Stosszahlansatz, or assumption of molecular chaos). Since the objections of Loschmidt and Zermelo were based solely on mechanical considerations, it is immediately obvious that the two paradoxes of the previous section are not really paradoxes at all. The only real problem, then, is to understand the precise nature of the Stosszahlansatz and how, in view of Poincaré cycles, a system can remain in equilibrium. One way out of Poincaré cycles is to say that they occupy an enormously long time for a large system [roughly of order $\exp(N)$ for a system of N particles]. Boltzmann, in fact, reputedly said to Poincaré: "You should wait that long." But this, of course, does not solve the problem.

Stated precisely now, Boltzmann's H theorem is as follows:

H THEOREM. *If at time τ a gas is in a state of molecular chaos* (i.e., a state satisfying the assumptions of molecular chaos), *then as $t \to \tau+$,*

(a) $dH/dt \leq 0$

and

(b) $dH/dt = 0$ *if and only if $f(\mathbf{v}, t)$ is the Maxwell–Boltzmann distribution, where*

$$H(t) = \int d^3v f(\mathbf{v}, t) \log f(\mathbf{v}, t)$$

and $f(\mathbf{v}, t)$ is a solution of Boltzmann's equation (4.15).

Recall also the definition of molecular chaos: If $f(\mathbf{v}, t)$ is the probability density for finding a molecule with velocity $\mathbf{v}$ at time t, then the probability density for simultaneously finding a molecule with velocity $\mathbf{v}_1$ and a molecule with velocity $\mathbf{v}_2$ at time t is $f(\mathbf{v}_1, t)f(\mathbf{v}_2, t)$.

It follows from the above statement and the results of the previous section that reversal and recurrence must take place when the state is not one of molecular chaos. The natural question to ask then is: When is a gas in a state of molecular chaos? The answer we feel must be "almost always," but it is clearly a hopeless task to actually prove it. We can nevertheless prove some results that support this view. The first result [attributed to F. E. Low by Huang (B1963)] is

1. *When a gas is in a state of molecular chaos, H is at a local peak.*

It is perhaps ironical that this result is an immediate consequence of time-reversal invariance.

Consider a gas without external forces in an initial state that is invariant under time reversal. The distribution function $f(\mathbf{v}, t)$ will then depend only on the magnitude of $\mathbf{v}$ and not on its direction. Let the gas be in a state of molecular chaos and be non-Maxwellian at time $t = 0$. Then according to the *H* theorem

$$\frac{dH}{dt} \leq 0 \qquad \text{at } t = 0+.$$

On the other hand, if we consider another gas with the same initial conditions but with velocities reversed, *H* is the same and

$$\frac{dH}{dt} \leq 0 \qquad \text{at } t = 0+.$$

But by time-reversal invariance, the future of the new gas is the past of the old. Therefore, for the original gas we have

$$\frac{dH}{dt} \leq 0 \qquad \text{at } t = 0+$$

and

$$\frac{dH}{dt} \geq 0 \qquad \text{at } t = 0-,$$

which proves statement 1.

Note that it is obvious from this result that when H is not at a local peak the gas is not in a state of molecular chaos, so Boltzmann's equation cannot be valid at all times. The above observation also allows us to make the following *time-reversal-invariant statement of the H theorem*:

2. *If there is molecular chaos at time* $t = \tau$, *then* $dH/dt \leq 0$ *at time* $t = \tau+$, *and if there is molecular chaos at time* $t = \tau+$, *then* $dH/dt \geq 0$ *at time* $t = \tau$.

The next observation (see Problem 4) is

3. *Independently of molecular chaos, H has its smallest value when the distribution is Maxwell–Boltzmann.*

We expect then that if collisions occur randomly, the distribution function will be almost always Maxwell–Boltzmann, and that fluctuations away from equilibrium will have a very short duration, typically of the order of 10^{-11} second, the time between two successive collisions.

In conclusion, then, we might argue (heuristically) that whenever the distribution function is not strictly Maxwell–Boltzmann, H is likely to be at a local peak, or, in other words, in a state of molecular chaos. Furthermore, if we imagine a gas to be initially in an improbable state we might expect H to look something like the curve shown in Figure 1.4, where the dashed curve represents the Boltzmann curve, the solid line the real curve, and the dotted peaks the states of molecular chaos. *In some sense, then, which is still not perfectly clear, the Boltzmann equation may be regarded as valid "on the average."*

The foregoing has been to a large extent descriptive. We consider in the next section a model that will serve to illustrate mathematically some of the ideas we have been discussing.

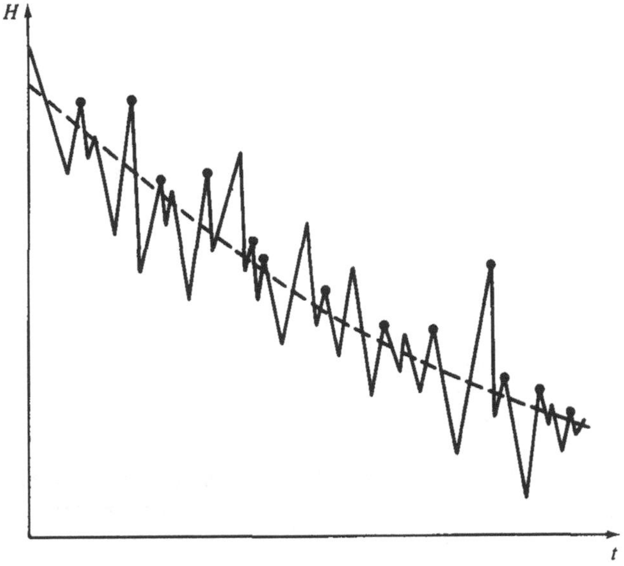

FIGURE 1.4. *H* as a function of time. The real curve is shown as a solid line and the dotted peaks represent states of molecular chaos. The dashed curve represents Boltzmann's *H* function.

1-9 The Kac Ring Model and the Stosszahlansatz

Consider a ring of n equally spaced points $P_1, P_2, \ldots, P_n$, m of which are marked (with a cross on Figure 1.5). Between successive points there is a ball that is either black or white, and during an elementary (unit) time interval each ball moves clockwise such that a ball changes color if and only if it moves through a marked point. The question is: Given a distribution of black and white balls at time $t = 0$, what is the distribution after time t? "Equilibrium" in this model corresponds to equal numbers of black and white balls, and, according to Boltzmann, equilibrium will be approached as a result of "collisions" (i.e., changing color).

This model was first proposed by Kac (1956). It has been discussed and generalized at length by Dresden (1962), and a nice discussion of the model can be found in Wannier (B1966). Similar and related models are discussed by Kac (B1959).

If we let $N_w(t)$ and $N_b(t)$, respectively, be the number of white and black

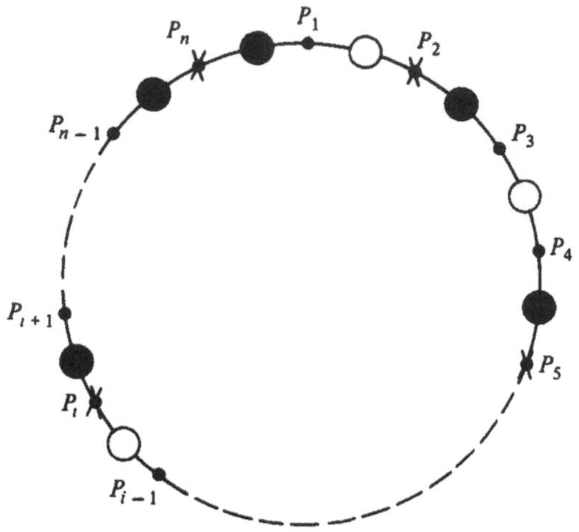

FIGURE 1.5. Kac ring model. Black and white balls move on a ring through points P_k, $k = 1, \ldots, n$ in unit time. If a ball moves through a marked point (crossed), it changes color.

balls at time t and $n_w(t)$ and $n_b(t)$ the number of white and black balls with a marked point ahead at time t, we have the *equations of motion*:

$$\begin{aligned} N_w(t+1) &= N_w(t) + n_b(t) - n_w(t), \\ N_b(t+1) &= N_b(t) + n_w(t) - n_b(t) \end{aligned} \tag{9.1}$$

and the *conservation equations*:

$$\left.\begin{aligned} N_w(t) + N_b(t) &= n \\ n_w(t) + n_b(t) &= m \end{aligned}\right\} \text{ for all time } t. \tag{9.2}$$

Equations 9.1 and 9.2 alone are clearly insufficient to determine $N_w(t)$ and $N_b(t)$, so another ingredient must be added. For example, Boltzmann's Stosszahlansatz for this model would state that the color of a ball is uncorrelated with the property of having a marked point ahead of it. In other words, $n_w(t)$ is proportional to $N_w(t)$ and $n_b(t)$ is proportional to $N_b(t)$. From Equations 9.2 this implies that

$$n_w(t) = \frac{m}{n} N_w(t)$$

and (9.3)

$$n_b(t) = \frac{m}{n} N_b(t).$$

With these relations Equations 9.1 give

$$\begin{aligned} N_b(t) - N_w(t) &= (1 - 2\mu)[N_b(t-1) - N_w(t-1)] \\ &= (1-2\mu)^t[N_b(0) - N_w(0)], \end{aligned} \tag{9.4}$$

where $\mu = m/n$.

Assuming now (and henceforth) that $2\mu < 1$, Equation 9.4 implies that equilibrium (i.e., $N_b = N_w$) is approached monotonically with time.

This conclusion is clearly untenable, since the model is completely *reversible and periodic*. Thus if after time t the balls are allowed to move counterclockwise, they will after time t return to their initial state, which is precluded by Equation 9.4.

It is obvious that the origin of the difficulty is Equation 9.3 which should be given some "statistical interpretation."

We now set up the model probabilistically and show that equilibrium is approached, essentially according to Equation 9.4 if one averages over distributions of marked points.

It is convenient first to introduce variables ε_j and $\eta_j(t)$ defined by

$$\varepsilon_j = \begin{cases} +1 & \text{if there is no marker at point } P_j \\ -1 & \text{if there is a marker at point } P_j \end{cases} \tag{9.5}$$

and

$$\eta_j(t) = \begin{cases} +1 & \text{if the ball between } P_{j-1} \text{ and } P_j \text{ is black at time } t \\ -1 & \text{if the ball between } P_{j-1} \text{ and } P_j \text{ is white at time } t. \end{cases} \tag{9.6}$$

It follows that

$$\eta_j(t) = \varepsilon_{j-1}\eta_{j-1}(t-1) \tag{9.7}$$

and hence that

$$\eta_j(t) = \varepsilon_{j-1}\varepsilon_{j-2}\cdots\varepsilon_{j-t}\,\eta_{j-t}(0), \tag{9.8}$$

where the indices are to be taken modulo n. By definition,

$$N_w(t) = \tfrac{1}{2}\sum_{\nu=1}^{n}[1 - \eta_\nu(t)] \tag{9.9}$$

and

$$N_b(t) = \tfrac{1}{2}\sum_{\nu=1}^{n}[1 + \eta_\nu(t)], \tag{9.10}$$

so, from Equation 9.8,

$$\begin{aligned} N_b(t) - N_w(t) &= \sum_{\nu=1}^{n} \eta_\nu(t) \\ &= \sum_{\nu=1}^{n} \varepsilon_{\nu-1}\varepsilon_{\nu-2} \cdots \varepsilon_{\nu-t}\, \eta_{\nu-t}(0). \end{aligned} \tag{9.11}$$

The distribution of markers determines the mechanical properties of the system, so no further progress can be made without knowledge of this distribution. Since we are not particularly interested in the exact distribution of markers, it is natural to perform an "average over markers." That is, if we denote probabilistic average by $\langle \cdots \rangle$, we have

$$\langle N_b(t) - N_w(t) \rangle = \sum_{\nu=1}^{n} \langle \varepsilon_{\nu-1}\varepsilon_{\nu-2} \cdots \varepsilon_{\nu-t} \rangle \eta_{\nu-t}(0). \tag{9.12}$$

If we now assume that the marker positions are equally probable, it follows that the average of any t consecutive factors ε_k is independent of the first factor, so, from Equations 9.11 and 9.12,

$$\langle N_b(t) - N_w(t) \rangle = \langle \varepsilon_1 \varepsilon_2 \cdots \varepsilon_t \rangle [N_b(0) - N_w(0)]. \tag{9.13}$$

For simplicity we now let $\mu = m/n$ be the probability that there is a marker at a point P_j independently of j. (Kac does not make this restriction; the final results are, however, the same.) If $t \le n$, the probability that there are s markers among t points is the binomial distribution

$$p(s, t) = \frac{t!}{s!(t-s)!} \mu^s (1-\mu)^t. \tag{9.14}$$

But if s of the t points $P_1, P_2, \ldots, P_t$ are marked

$$\varepsilon_1 \varepsilon_2 \cdots \varepsilon_t = (-1)^s; \tag{9.15}$$

hence

$$\begin{aligned} \langle \varepsilon_1 \varepsilon_2 \cdots \varepsilon_t \rangle &= \sum_{s=0}^{t} (-1)^s p(s, t) \\ &= (1 - 2\mu)^t \qquad \text{provided } t \le n. \end{aligned} \tag{9.16}$$

The case $t \ge n$ is left as an exercise [see Wannier (B1966)]. For our purposes we now keep t fixed and let n and $m \to \infty$ with $\mu = m/n$ fixed. From Equations 9.13 and 9.16 we then obtain

$$\lim_{\substack{n, m \to \infty \\ \mu \text{ fixed}}} \langle N_b(t) - N_w(t) \rangle = (1 - 2\mu)^t [N_b(0) - N_w(0)]. \tag{9.17}$$

Equilibrium is then approached monotonically as before in the limit $t \to \infty$.

It is important to note that two processes are needed here to prepare the system for an approach to equilibrium: an average over markers and the limit of an infinite system.

This model illustrates well the mathematical origins of irreversibility, which are definitely not mechanical, and in the present instance result in a sense from an average over "disturbances." It should be stressed that the type of averaging and the limiting process necessary to get Boltzmann's result here is not easily generalized to bona fide mechanical systems. Luttinger and Kohn (1958) used similar trickery to discuss the scattering of electrons by fixed impurities with random locations in a solid, but for gases one is hard put to find something that can be randomized as readily. Gibbs suggested that an average over initial conditions may perform that function. This does not seem to work for the Kac model but may conceivably work in some cases.

The correct view in the case of gases may well be that statistics is never done on a mechanical sytem that is completely shielded from outside influences. That is, the "Hamiltonian" does not describe the system completely, and stray interactions from "outside" randomize certain features that would otherwise remain fixed from mechanics.

In general, the relation between "averaging" and the Stosszahlansatz is unclear, and this remains one of the major difficulties with the theory.

1-10 Concluding Remarks

At this point we leave kinetic theory and move on to thermodynamics and Gibbsian statistical mechanics, which from the mathematical point of view are somewhat more appealing. The problems of equilibrium statistical mechanics can be well formulated mathematically, and there is general agreement among physicists on the correctness of the basic axioms of the theory. The situation is, as we have seen, a little different in kinetic theory or, if you like, nonequilibrium statistical mechanics.

Apart from the problem of actually solving equations such as Boltzmann's equation (and this has become an art in itself!) there is the fundamental problem, still not completely resolved, of understanding the relation between the irreversible "statistical equations" and the reversible "dynamical equations." The problem really is when and how to introduce statistics.

Most of the recent research in this area has been concerned with "deriving" the Boltzmann equation from the Liouville equation, and the corrections resulting from three-body, four-body, etc., collisions. Developments of this type have led to "density expansions" for various transport coefficients (such as viscosity), but it has been shown recently by Cohen and others [e.g., Dorfman and Cohen (1967)] that such expansions do not exist. The first-order corrections are all right, but after that logarithmically divergent terms begin to appear. So we are back where we started, with the question asked last century: In what sense is Boltzmann's equation a valid first approximation?

PROBLEMS

1. Assuming that the atoms of a gas satisfy Maxwell's distribution, evaluate the number of atoms that strike a unit area in unit time.

2. Calculate the mean-square fluctuation of velocity $\overline{(v-\bar{v})^2}$ for the Maxwell gas, where

$$\overline{v^n} = \left(\frac{m}{2\pi kT}\right)^{3/2} \iiint_{-\infty}^{\infty} \exp\left(-\frac{mv^2}{2kT}\right) v^n \, dv_1 \, dv_2 \, dv_3$$

and

$$v^2 = v_1^2 + v_2^2 + v_3^2. \left[\text{Note that } \int_{-\infty}^{\infty} \exp(-x^2)\, dx = \pi^{1/2}.\right]$$

3. During an elastic collision of two (equal-mass) hard spheres, energy and momentum are conserved, i.e.,

$$v_1^2 + v_2^2 = v_1'^2 + v_2'^2$$

and

$$\mathbf{v}_1 + \mathbf{v}_2 = \mathbf{v}_1' + \mathbf{v}_2',$$

where $(\mathbf{v}_1, \mathbf{v}_2)$ and $(\mathbf{v}_1', \mathbf{v}_2')$ are the initial and final velocities, respectively, of the two molecules.

Deduce that if $\mathbf{g} = \mathbf{v}_2 - \mathbf{v}_1$ and $\mathbf{g}' = \mathbf{v}_2' - \mathbf{v}_1'$,

(a) $|\mathbf{g}| = |\mathbf{g}'|$;

(b) $\mathbf{v}_1' = \mathbf{v}_1 + [(\mathbf{v}_2 - \mathbf{v}_1) \cdot \mathbf{l}]\mathbf{l}$

and

$$\mathbf{v}_2' = \mathbf{v}_2 - [(\mathbf{v}_2 - \mathbf{v}_1) \cdot \mathbf{l}]\mathbf{l},$$

where $\mathbf{l}$ is the unit vector in the direction of the line of centers of the two spheres at the time of collision. [*Hint:* Show that the components of $\mathbf{g}$ and $\mathbf{g}'$ parallel to $\mathbf{l}$ are equal in magnitude and opposite in sign and that the components perpendicular to $\mathbf{l}$ are equal in magnitude and sign. In other words, $\mathbf{g} - \mathbf{g}' = 2(\mathbf{g} \cdot \mathbf{l})\mathbf{l}$.]

(c) $d^3v_1\, d^3v_2 = d^3v_1'\, d^3v_2'$, or the transformation $(\mathbf{v}_1, \mathbf{v}_2) \rightarrow (\mathbf{v}_1', \mathbf{v}_2')$ is orthogonal.

4. Let $H(t) = \int d^3v f(\mathbf{v}, t) \log f(\mathbf{v}, t)$, where $f(\mathbf{v}, t)$ is arbitrary except for the conditions (1) $\int d^3v f(\mathbf{v}, t) = N/V$ and (2) $\int d^3v f(\mathbf{v}, t) mv^2/2 = \varepsilon$. Show that H is a minimum when $f(\mathbf{v}, t)$ is the Maxwell–Boltzmann distribution.

5. Show that if

$$f(\mathbf{v}, \mathbf{r}) = f_0(\mathbf{v}) \exp\left[-\frac{\phi(\mathbf{r})}{kT}\right]$$

where $f_0(\mathbf{v})$ is Maxwell's distribution and

$$\mathbf{F} = -\nabla_\mathbf{r} \phi(\mathbf{r}),$$

then

$$\left(\mathbf{v} \cdot \nabla_\mathbf{r} + \frac{\mathbf{F}}{m} \cdot \nabla_\mathbf{v}\right) f(\mathbf{v}, \mathbf{r}) = 0.$$

6. A vessel of volume V contains N molecules. Let n be the number of molecules in a part of the vessel with volume v, and in thermal equilibrium let $p = v/V$ be the probability that a given molecule is in v.

(a) Show that the probability distribution $f(n)$ for the number n is the binomial distribution

$$\frac{N!}{n!(N-n)!} p^n (1-p)^{N-n}.$$

(b) From the generating function $G(x) = \sum_{n=0}^{\infty} f(n) x^n$ or otherwise, find the mean number of molecules $\bar{n}$ and the mean-square fluctuation $\overline{(n - \bar{n})^2}$ of the number of molecules in v.

(c) Making use of Stirling's formula ($\log N! \sim N \log N - N$ as $N \to \infty$), show that

$$f(n) \sim A \exp\left[-\frac{(n-\bar{n})^2}{2\Delta}\right] \qquad \text{as } n, N \to \infty \ (n/N \text{ fixed}).$$

(d) Show that in the limit $N, V \to \infty$ with N/V and v fixed, $f(n)$ approaches the Poisson distribution

$$\frac{e^{-\bar{n}}(\bar{n})^n}{n!}$$

[*Hint:* Investigate the limiting behavior of the generating function $G(x)$.]

7. Consider the following ("wind-tree") model of P. and T. Ehrenfest [Ehrenfest and Ehrenfest (B1911)]. In a plane there are N "P molecules" per unit area. The P molecules do not interact with one another but collide elastically with "Q molecules," which are fixed squares with side a distributed at random over the plane with diagonals parallel to the x and y axes. Their average surface density is n and let their mean distance apart be large compared with a.

Suppose that at a certain moment all the P molecules have constant velocities c in any of the four coordinate directions. Owing to the nature of the allowed interactions, this situation will hold at any other time except possibly for the number of P molecules moving in the four coordinate directions. Let f_1, f_2, f_3, and f_4 be the numbers of P molecules moving in those four directions. These will be functions of time t and together take the place of the distribution function $f(\mathbf{v}, t)$. The equilibrium distribution is clearly

$$f_1 = f_2 = f_3 = f_4 = \frac{N}{4}$$

and the problem is to show that equilibrium is approached monotonically, assuming that the P molecules obey the Stosszahlansatz

$$N_{ij}\,\delta t = f_i\,Sn,$$

where $N_{ij}\,\delta t$ is the number of P molecules per unit area changing direction from i to j during time δt and S is the area of the parallelogram shown in Figure 1.6 (for $i = 1, j = 2$), i.e., $S = c\,\delta t a/\sqrt{2}$.

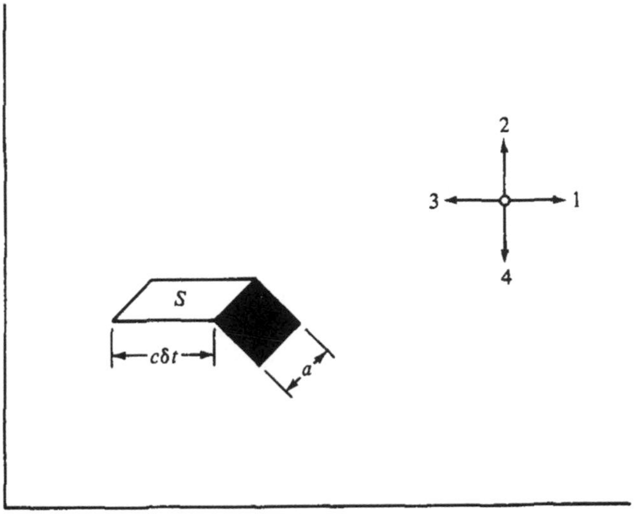

FIGURE 1.6. Ehrenfest wind-tree model.

CHAPTER 2

Thermodynamics

2-1 Introduction

The main aim of statistical mechanics is to derive *macroscopic* properties of matter from the laws governing the *microscopic* interactions of the individual particles.

Thermodynamics consists of laws governing the behavior of macroscopic variables and as such is a necessary prerequisite for statistical mechanics. It is a subject in its own right and has the advantage, from the mathematical point of view, that it can be completely axiomatized. Since our main concern here, however, is statistical mechanics, we shall present only the "primitive concepts" and "laws" and, for later reference, some particular thermodynamic relations. A detailed account of the axiomatization of thermodynamics [due mainly to Carathéodory (1909)] can be found, for example, in Buchdahl (B1966). More extensive treatments of the subject than ours are Callen (B1960) and Pippard (B1957).

The primitive concepts are obvious undefined things, analogous to "point" and "line" in Euclidean geometry, such as

1. "System": for example a gas.
2. "State of a system": specified, for example, by pressure P and volume V.
3. "Thermal equilibrium of a system," which is a state of a system that does not change with time.

4. "Equation of state": for example, an equation relating P, V, and temperature T, for example $PV = NkT$ for an ideal gas.

and so forth. It should be clear in the following sections which concepts are primitive and which are not.

We have discussed the notions of state and equilibrium in Chapter 1, but there the problem was to show, for example, that equilibrium can result from collisions. Thermodynamics is merely concerned with laws governing the behavior of macroscopic (thermodynamic) variables.

It is not really surprising that thermodynamics was developed at about the same time as kinetic theory. In fact, a number of people, notably Clausius, made significant contributions to both subjects. Some historical remarks are given at the end of Section 2-3.

We now discuss the various laws of thermodynamics.

2-2 Empirical Temperature and the Zeroth Law

As the name suggests, the zeroth law was added as an afterthought. Its main point is to define thermal equilibrium and to introduce the notion of temperature.

Here and henceforth we shall consider only one-component systems whose states are specified by a value of pressure P and a value of volume V. In other words, a state of the system is specified by a point in the first quadrant of the (P, V) plane. Note that P and V are appropriate variables for a gas. For a magnet, we would take magnetization M and magnetic field H, which are analogous to P and V, respectively. More complicated systems, whose states are specified by more than two variables, can be handled in a similar fashion. [See, for example, Callen (B1960).]

We now introduce the notion of equilibrium of two systems through the following

POSTULATE. *Associated with each pair of systems A and B there is a function $f_{AB}(P_1, V_1; P_2, V_2)$ of the states (P_1, V_1) of A and (P_2, V_2) of B such that A and B are in equilibrium if and only if*

$$f_{AB}(P_1, V_1; P_2, V_2) = 0. \tag{2.1}$$

Note that this postulate implies that the states of two systems in equilibrium cannot be specified arbitrarily.

We can now state the

ZEROTH LAW OF THERMODYNAMICS. *A state of equilibrium exists and is transitive. That is, if a system A is in equilibrium with a system B and B is in equilibrium with a system C, then A is in equilibrium with C.*

In view of Equation 2.1, transitivity means that

$$f_{AB}(P_1, V_1; P_2, V_2) = 0 \qquad \text{and} \qquad f_{BC}(P_2, V_2; P_3, V_3) = 0$$

imply that (2.2)

$$f_{AC}(P_1, V_1; P_3, V_3) = 0.$$

Transitivity allows us to define a temperature scale, specifically *empirical temperature*, associated with a given system. An absolute scale of temperature will follow later as a consequence of the first and second laws.

To derive the notion of temperature one can argue in a number of ways. The following is perhaps the simplest. Consider two systems A and C in equilibrium with a "test system" (or "thermometer") B. We assume that for a test system the first two equations of Equation 2.2 can be solved for P_2 in terms of P_1, V_1, V_2 and P_3, V_3, V_2, respectively (or V_2 in terms of P_1, V_1, P_2 and P_3, V_3, P_2, respectively). From the implicit-function theorem, this means that $f_{XB}(P_1, V_1; P_2, V_2)$ (X being A or C in the present instance) must be continuous with continuous first partial derivatives and that $\partial f_{XB}/\partial P_2$ must not vanish. Granted the existence of such a test system B, there then exist functions Θ_A and Θ_C such that

$$\Theta_A(P_1, V_1; V_2) = \Theta_C(P_3, V_3; V_2), \tag{2.3}$$

which is equivalent, from the zeroth law,. to saying that A and C are in equilibrium.

Now, since B is a fixed system, we can ignore the V_2 dependence of Θ_A and Θ_C. $\Theta_A(P_1, V_1)$ is then defined to be the empirical temperature of the system A, *as measured by the thermometer B*. A necessary and sufficient condition then for two systems to be in equilibrium is that they have the same empirical temperature (as measured by a test system or thermometer).

It is important to note that equilibrium via a test system is necessary for a proper definition of empirical temperature. A common error in a number of books states that since Equation 2.3, written as

$$\Theta_A(P_1, V_1; V_2) - \Theta_C(P_3, V_3; V_2) = 0, \tag{2.4}$$

is equivalent to $f_{AC}(P_1, V_1; P_3, V_3) = 0$ (Equation 2.2), the left-hand side of

Equation 2.4 must be independent of V_2 and hence Θ_A and Θ_C must be of the form

$$\Theta_A(P_1, V_1; V_2) = \theta_A(P_1, V_1) + \psi(V_2)$$

and (2.5)

$$\Theta_C(P_3, V_3; V_2) = \theta_C(P_3, V_3) + \psi(V_2).$$

$\theta_A(P_1, V_1)$ is then defined to be the empirical temperature of A. The error, of course, is that Θ_A and Θ_C need not have the form of Equation 2.5!

Note that if $\theta = \Theta(P, V)$ is the empirical temperature of a system, and if we can invert the function Θ to obtain $P = P(V, \theta)$, we have the equation of state for the system (i.e., pressure as a function of volume and temperature). One of the aims of statistical mechanics is to derive an analytic expression for the equation of state. This is clearly beyond the scope of thermodynamics.

With the notion of temperature we now move on to the first law.

2-3 The First Law

Before stating the first law, it is necessary to define some terms. The states (P, V) of a system A with constant empirical temperature $\Theta_A(P, V)$ form a family of curves in the (P, V) plane called *isotherms*.

It follows that through each point in the (P, V) plane there is one and only one isotherm.

A *quasi-static process* is a change of state that takes place so slowly that the system is almost in thermal equilibrium (with a test system) at each step of the process.

A process is *reversible* if the initial state can be obtained from the final state by following the inverse of each step of the original process.

If a gas changes quasi-statically from state 1 to state 2 along a curve Γ in the (P, V) plane as shown in Figure 2.1, the work done by the system in the process is defined to be

$$\Delta W = \int_\Gamma P\,dV. \tag{3.1}$$

Note that the above change of state is reversible if state 1 can be reached from state 2 along the path $-\Gamma$.

Finally, an *adiabatic process* is a change of state that occurs while the system

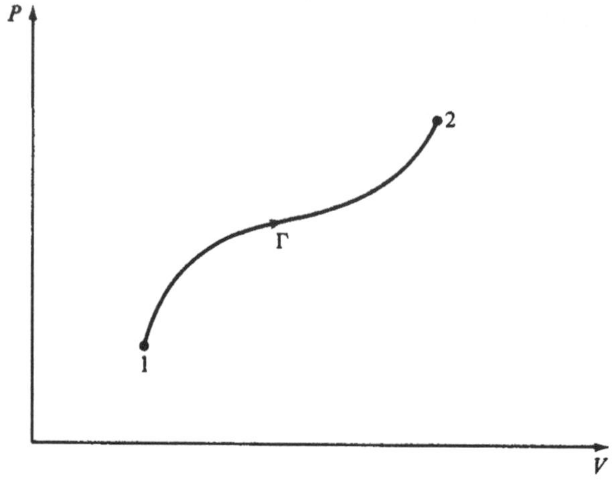

FIGURE 2.1. Quasi-static change from state 1 to state 2 along a path Γ.

is *thermally isolated* (i.e., not in contact or in equilibrium with any other system).

We can now state the

FIRST LAW OF THERMODYNAMICS.

1. *Associated with each system A there is a function* $U_A(P, V)$, *called the* internal energy *of A, such that the work* ΔW *done by the system in going from one state* (P_1, V_1) *to another state* (P_2, V_2) *in any* adiabatic process *is*

$$\Delta W = U_A(P_1, V_1) - U_A(P_2, V_2) = -\Delta U_A \tag{3.2}$$

independently of the path joining the initial and final states. It follows from definition 3.1 *that if the process is also quasi-static,* $\int_\Gamma P\,dV$ *is independent of the path* Γ.

2. *The* amount of heat Q_A *absorbed by the system in* any *change of state is defined by*

$$Q_A = \Delta U_A + \Delta W \tag{3.3}$$

and for an infinitesimal quasi-static change

$$\delta Q = dU + P\,dV, \tag{3.4}$$

where we have written δQ, *since* $dU + P\,dV$ *is not necessarily an exact differential of some function.*

For quasi-static processes,

$$U(P_2, V_2) - U(P_1, V_1) = \int_\Gamma dU = \int_\Gamma (\delta Q - P\,dV), \tag{3.5}$$

where Γ is any path joining the initial state (P_1, V_1) and the final state (P_2, V_2). It follows that in any quasi-static cyclic process, i.e., with same initial and final states, there is no change in the internal energy of the system even though heat may be absorbed and work may be done by the system. Clearly this is just a statement of *energy conservation.*

It is interesting to trace the history of heat and energy conservation to Clausius' statement [Clausius (1850)] of the first and second laws of thermodynamics. Heat, as everyone now knows, is a form of energy, but even as late as the middle of the last century it was almost universally considered as a substance, commonly known as caloric, and definitely not as a form of energy. Rumford (1798) was probably the first to attempt an experiment (using cannons and horses!) to show that mechanical energy can be transformed into heat. His experiment simply showed that when a cannon was bored in water, the water temperature increased. Davy (1799) performed similar experiments with ice, but since neither Rumford nor Davy had a satisfactory alternative theory, the caloric theory survived. More decisive experiments by Mayer in 1842 and especially by Joule (1845) paved the way for Clausius' "mechanical theory of heat" [Clausius (1850)], in which the first and second laws are stated.

The second law, which is discussed in the next section, was actually formulated earlier by Carnot (1824). His arguments were based in part on the caloric theory, however, so his work was neglected somewhat by later workers.

The reader interested in more historical details is referred to Roller (B1950) and Brown (1950).

2-4 Carnot Cycles and the Second Law

The second law can be discussed in a number of different but essentially equivalent ways. We shall proceed in a more or less standard and historical way via Carnot cycles, Kelvin's statement [which is equivalent to Clausius' statement (Problem 1)], the absolute-temperature scale, and entropy.

We begin by defining some terms. Consider a quasi-static adiabatic process, i.e., a process such that

$$\delta Q = dU + P\,dV = 0,$$

or, equivalently, using the fact that $U = U(P, V)$,

$$\left(P + \frac{\partial U}{\partial V}\right) dV + \frac{\partial U}{\partial P}\, dP = 0. \tag{4.1}$$

The solutions of the first-order differential equation (4.1) are called *adiabatic curves.*

We *assume* that $U(P, V)$ is sufficiently regular so that there is one and only one adiabatic curve passing through a given point (P_0, V_0). In other words, given some state (P_0, V_0) we can find another state (P', V') which can be connected to (P_0, V_0) by an adiabatic curve.

An object that we shall frequently use is a *heat bath,* which is defined to be a system so large that its empirical temperature remains constant when a finite amount of heat is added to it or subtracted from it.

We now define a *Carnot cycle* for a particular system to be the *reversible* cyclic process shown schematically in Figure 2.2.

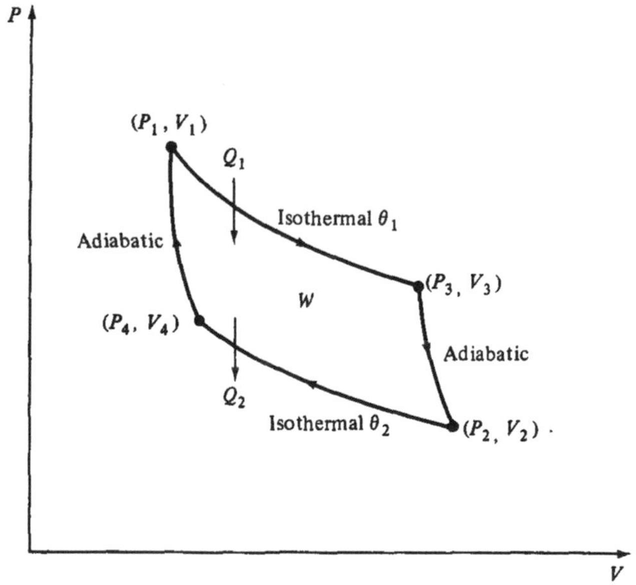

FIGURE 2.2. Carnot cycle (see the text).

Specifically, we start from a state (P_1, V_1) with empirical temperature $\theta_1 = \Theta(P_1, V_1)$ and expand the system isothermally to (P_3, V_3), keeping it in thermal equilibrium with a heat bath at temperature θ_1. In this process a certain amount of heat Q_1 is absorbed by the system (from the heat bath). Next we expand the system adiabatically to (P_2, V_2), at which point the system has temperature $\theta_2 = \Theta(P_2, V_2)$. We then compress the system isothermally while it is in thermal equilibrium with a heat bath at temperature θ_2 until it reaches a state (P_4, V_4), which can then be compressed adiabatically back to the initial state (P_1, V_1). During the isothermal compression, an amount of heat Q_2 is absorbed by the heat bath and, assuming that all processes are performed quasi-statically, the first law implies that

$$Q_1 = \int_{V_1}^{V_3} P(V, \theta_1)\, dV + E(V_3, \theta_1) - E(V_1, \theta_1) \tag{4.2a}$$

and

$$Q_2 = \int_{V_4}^{V_2} P(V, \theta_2)\, dV + E(V_2, \theta_2) - E(V_4, \theta_2) \tag{4.2b}$$

along the two isotherms and that

$$0 = \int_{V_3}^{V_2} P\, dV + E(V_2, \theta_2) - E(V_3, \theta_1) \tag{4.2c}$$

and

$$0 = \int_{V_1}^{V_4} P\, dV + E(V_4, \theta_2) - E(V_1, \theta_1) \tag{4.2d}$$

along the two adiabatic portions of the cycle. In Equations 4.2 we have used the relation $\theta = \Theta(P, V)$ to express the pressure as $P = P(V, \theta)$ and the internal energy as

$$U(P, V) = U(P(V, \theta), V) = E(V, \theta). \tag{4.3}$$

Denoting the Carnot-cycle curve by C we deduce from Equations 4.2 that the total work W done by the gas during the cycle is

$$W = \oint_C P\, dV = Q_1 - Q_2. \tag{4.4}$$

We define now the efficiency e of the process described above to be the ratio of the work done by the gas to the quantity of heat taken in during the process, i.e.,

$$e = \frac{W}{Q_1} = 1 - \frac{Q_2}{Q_1}. \tag{4.5}$$

We will show in a moment that the second law of thermodynamics implies that e is independent of the particular system undergoing the process and that as a result an absolute scale of temperature can be defined.

We start with

KELVIN'S STATEMENT OF THE SECOND LAW. *No cyclic process exists whose sole effect is to extract heat from a substance and convert it entirely to work.* In other words, there can be no perpetual-motion machine which converts the whole internal energy of a body into mechanical work. (An equivalent statement due to Clausius is given in Problem 1.)

Physically speaking, this statement is eminently reasonable. It means, for example, that pencils (or anything else) are not allowed to convert their internal energy into potential energy by suddenly jumping up from a table.

We shall now deduce from Kelvin's statement that the efficiency e of a Carnot cycle depends only on θ_1 and θ_2 and not on the particular system undergoing the cycle.

Consider two systems A and A' undergoing Carnot cycles between temperatures θ_1 and θ_2. We assume that the relative size of A and A' is such that $Q_1 = Q_1'$ (this is not necessary, but it simplifies the argument). We now show that $Q_2 = Q_2'$ or, equivalently, $W = W'$. Suppose that this is not so and that $W' > W$ or, equivalently, $Q_2' < Q_2$. Consider the combined process obtained by running the Carnot cycle for A' followed by the *reversed* Carnot cycle for A. This process absorbs a positive amount of heat $Q_2 - Q_2'$ at temperature θ_2 and converts it entirely into work. This contradicts Kelvin's statement; hence $Q_2' \nless Q_2$. Similarly, $Q_2 \ngtr Q_2'$; hence $Q_2 = Q_2'$ and the assertion is proved.

It follows immediately that the ratio

$$\frac{Q_2}{Q_1} = f(\theta_1, \theta_2) \tag{4.6}$$

is a universal function of θ_1 and θ_2. We now show that $f(\theta_1, \theta_2)$ must in fact be of the form

$$f(\theta_1, \theta_2) = \frac{\Phi(\theta_2)}{\Phi(\theta_1)}. \tag{4.7}$$

To prove this we consider two Carnot cycles. The first operates between temperatures θ_1 and θ_2, absorbing a quantity of heat Q_1 during the first isothermal change and yielding a quantity of heat Q_2 during the second isothermal change. The second cycle operates between temperatures θ_2 and θ_3, absorbing a quantity of heat Q_2 and yielding a quantity of heat Q_3. For these two cycles Equation 4.6 gives

$$\frac{Q_2}{Q_1} = f(\theta_1, \theta_2) \tag{4.8a}$$

and

$$\frac{Q_3}{Q_2} = f(\theta_2, \theta_3). \tag{4.8b}$$

For the combined cycle

$$\frac{Q_3}{Q_1} = f(\theta_1, \theta_3). \tag{4.8c}$$

Multiplying Equations 4.8a and 4.8b then gives, from Equation 4.8c,

$$f(\theta_1, \theta_3) = f(\theta_1, \theta_2) f(\theta_2, \theta_3), \tag{4.8d}$$

from which it follows that

$$f(\theta_1, \theta_2) = \frac{\Phi(\theta_2)}{\Phi(\theta_1)} \tag{4.9}$$

for some (universal) function Φ. $\Phi(\theta)$, by definition, is a universal function of empirical temperature θ. The *absolute temperature* T defined by

$$T = \alpha\Phi(\theta), \tag{4.10}$$

where α (a constant) is any scale factor, is then *universal.* It follows that the efficiency of a Carnot cycle for any gas operating between absolute temperatures T_1 and T_2 is given by

$$e = 1 - \frac{Q_2}{Q_1} = 1 - \frac{T_2}{T_1}. \tag{4.11}$$

Equation 4.11 may be rewritten in the form

$$\frac{Q_1}{T_1} + \frac{-Q_2}{T_2} = 0, \tag{4.12}$$

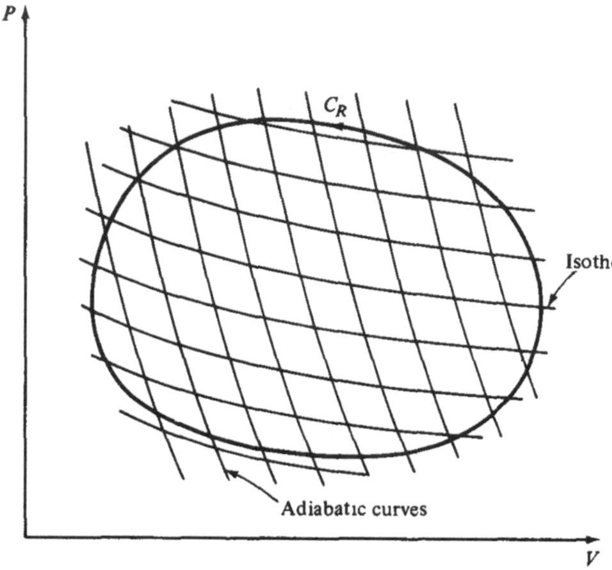

FIGURE 2.3. Approximation of a closed reversible cycle C_R by a network of Carnot cycles.

which says that the sum of (possibly negative) amounts of heat absorbed by a system, divided by the absolute temperatures at which they are absorbed, is zero. Consider now *any arbitrary closed reversible cycle* C_R approximated by a network of Carnot cycles as shown in Figure 2.3. Applying Equation 4.12 to each little Carnot cycle in the approximating network and allowing the network to become finer and finer, we obtain in the limit

CARNOT'S THEOREM. *For any closed reversible cycle* C_R,

$$\oint_{C_R} \frac{\delta Q}{T} = 0. \tag{4.13}$$

Carnot's theorem allows us to define a function of the state of a system (a state function, for short)

$$S = S(V, T) \tag{4.14}$$

called the *entropy*, by

$$dS = \frac{\delta Q}{T}. \tag{4.15}$$

Equation 4.13 implies that $\delta Q/T$ is an exact differential (of S). It is clear that S defined in this way is only determined up to an additive constant. In other words, only the difference in entropy between two states 1 and 2 defined by

$$S(V_2, T_2) - S(V_1, T_1) = \int_\Gamma \frac{\delta Q}{T} \tag{4.16}$$

is determined, where Γ is any *reversible* path joining states 1 and 2.

Let us consider now what happens to the above results if the paths in question are *not reversible*. In particular, consider any cycle C operating between two heat baths at absolute temperatures T_1 and T_2 absorbing a quantity of heat Q_1 from the first reservoir and a quantity of heat $-Q_2$ from the second reservoir (i.e., C is the "not-necessarily-reversible" cycle shown in Figure 2.2). In addition, consider two Carnot cycles C_0 and C_0' operating between temperatures T_0 and T_1 and T_0 and T_2, respectively, absorbing quantities of heat Q_0 and Q_0' at T_0 and yielding quantities of heat Q_1 and $-Q_2$ at T_1 and T_2, respectively. From Equation 4.11 we have

$$\frac{Q_0}{Q_1} = \frac{T_0}{T_1} \qquad \text{for cycle } C_0 \tag{4.17}$$

and

$$\frac{Q_0'}{-Q_2} = \frac{T_0}{T_2} \qquad \text{for cycle } C_0'. \tag{4.18}$$

Consider now the combined cycle consisting of C followed by C_0 followed by C_0'. A net amount of heat

$$Q = Q_0 + Q_0' \tag{4.19}$$

is absorbed by this system and is converted entirely into work. In order to avoid a contradiction with the second law, it then follows that

$$Q \leq 0; \tag{4.20}$$

in other words, from Equations 4.17 and 4.18 ($T_0 > 0$),

$$\frac{Q_1}{T_1} + \frac{-Q_2}{T_2} \leq 0. \tag{4.21}$$

Equation 4.21 is to be compared with Equation 4.12, which holds for reversible processes. In fact, if the cycle C is reversible, we can reverse the combined cycle $C + C_0 + C_0'$ to get Equation 4.21 with the inequality reversed. Equation 4.12 then follows.

We can now repeat the argument leading from Equation 4.12 to Equation 4.13 to obtain

CLAUSIUS' THEOREM. *For any cycle C*

$$\oint_C \frac{\delta Q}{T} \leq 0 \tag{4.22}$$

with equality holding if the cycle is reversible.

As a corollary to Clausius' theorem, we have (see Equation 4.16)

$$\int_\Gamma \frac{\delta Q}{T} \leq S(V_2, T_2) - S(V_1, T_1) \tag{4.23}$$

for any path Γ from state (V_1, T_1) to state (V_2, T_2), with equality holding if Γ is reversible. To prove this result, consider the cycle C made up of Γ and $-\Gamma_R$ as shown in Figure 2.4, with Γ_R a reversible path from state (V_1, T_1) to state (V_2, T_2). Equation 4.16 states that

$$S(V_2, T_2) - S(V_1, T_1) = \int_{\Gamma_R} \frac{\delta Q}{T}. \tag{4.24}$$

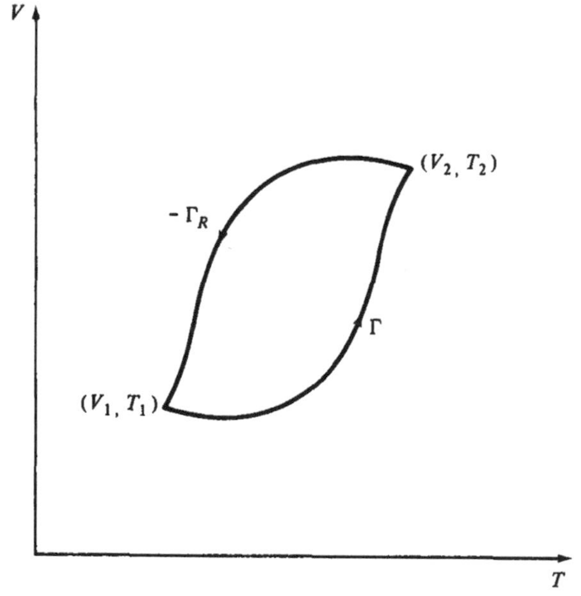

FIGURE 2.4. Reversible path $-\Gamma_R$ and nonreversible path Γ connecting states 1 and 2.

But, from Equation 4.22,

$$\int_{\Gamma} \frac{\delta Q}{T} - \int_{\Gamma_R} \frac{\delta Q}{T} \le 0. \tag{4.25}$$

Equation 4.23 then follows from Equations 4.24 and 4.25.

From Equation 4.23 we immediately deduce that for a thermally isolated system (i.e., $\delta Q = 0$ for any process), entropy never decreases.

We summarize the foregoing as an

ALTERNATIVE STATEMENT OF THE SECOND LAW:

1. *For* reversible *changes of state, there exists an absolute scale of temperature T such that $\delta Q/T$ is an exact differential of a function S, called entropy, of the state of the system.*

2. *For* irreversible *changes of state in a thermally isolated system, the entropy increases.*

There is one more "law" which came sometime after the first and second laws. It is sometimes called the *third law of thermodynamics* and sometimes *Nernst's theorem,* after Nernst, who first stated it in 1905. The statement is simply: *The entropy of a system at absolute zero is a universal constant which may be taken to be zero.*

Since we shall not be particularly concerned with low-temperature statistical mechanics or with such questions as the attainability of absolute zero, we shall not discuss this law here [the interested reader is referred to Huang (B1963) or any standard textbook on thermodynamics and statistical mechanics]. Instead we move on to thermodynamic potentials, which we shall need when we get into statistical mechanics proper.

2-5 Thermodynamic Potentials

The following two state functions, Ψ and Φ, are useful in determining the equilibrium state of a system: the *Helmholtz free energy* Ψ, defined by

$$\Psi = U - TS, \tag{5.1}$$

and the *Gibbs potential* Φ, defined by

$$\Phi = U - TS + PV. \tag{5.2}$$

The essential properties of Ψ and Φ are contained in the following

THEOREM:

1. *For a mechanically isolated system at constant temperature,* Ψ *never increases. The thermal equilibrium state for such a system is then the state of minimum* Ψ.

2. *For a system at constant pressure and temperature,* Φ *never increases. The equilibrium state for such a system is then the state of minimum* Φ.

PROOF:

1. From Equation 4.23 it follows that for an infinitesimal isothermal transformation Δ,

$$\frac{\Delta Q}{T} \leq \Delta S. \tag{5.3}$$

Now from the first-law Equation 3.3, we can rewrite Equation 5.3 as

$$\Delta W \leq -\Delta U + T\,\Delta S. \tag{5.4}$$

The right-hand side of Equation 5.4 is $-\Delta\Psi$, and for a mechanically isolated system $\Delta W = 0$. It follows that

$$\Delta\Psi \leq 0 \tag{5.5}$$

with equality holding if the transformation Δ is reversible.

2. For a system with constant pressure and temperature, Equation 5.4 holds and in addition

$$\Delta W = P\,\Delta V. \tag{5.6}$$

Equations 5.4 and 5.2 then imply that

$$\Delta\Phi = P\,\Delta V + \Delta U - T\,\Delta S \leq 0, \tag{5.7}$$

which completes the proof of the theorem.

We can consider Ψ or Φ as functions of any two of the three variables P, V, T, since these variables depend on one another through the equation of state. For example, if we consider, as is usual, $\Psi = \Psi(V, T)$, we find from Equation 5.1 that

$$\begin{aligned} d\Psi &= dU - T\,dS - S\,dT \\ &= \delta Q - P\,dV - T\,dS - S\,dT \\ &= -P\,dV - S\,dT. \end{aligned} \tag{5.8}$$

It follows immediately that

$$P = -\left(\frac{\partial \Psi}{\partial V}\right)_T \tag{5.9}$$

and

$$S = -\left(\frac{\partial \Psi}{\partial T}\right)_V, \tag{5.10}$$

where to be specific we have indicated by a subscript the variable to be held fixed during the differentiation. Equations 5.9 and 5.10 are two of a class of relations generally known as *Maxwell relations.* More are given as Problems at the end of the chapter.

The task of statistical mechanics is to actually derive analytic expressions for Ψ or Φ (or any other appropriate potential). Quantities of physical interest are then obtained straightforwardly by differentiation. For example, the pressure is given by Equation 5.9, from which one can then obtain the *isothermal compressibility* defined by

$$K_T = -\frac{1}{V}\left(\frac{\partial V}{\partial P}\right)_T, \tag{5.11}$$

and so forth.

One final point worth noting is that on physical grounds K_T must be non-negative. From Equations 5.9 and 5.11 this implies that

$$\left(\frac{\partial^2 \Psi}{\partial V^2}\right)_T \geq 0. \tag{5.12}$$

In other words, Ψ must be a *convex function* of volume (assuming that the second derivative exists), as shown in Figure 2.5.

Similarly, requiring that the specific heat at constant volume C_V be non-negative implies that Ψ as a function of temperature T must be a *concave* or *convex downward* function, as shown in Figure 2.6. (See Problem 4.)

These convexity properties of the free energy turn out to be rather important and will be discussed in more detail in later chapters.

2-6 The Ideal Gas and Some Thermodynamic Relations

Experimentally there are a number of systems that have an equation of state of the form

$$PV = \phi(T) \qquad \text{(Boyle's law)} \tag{6.1}$$

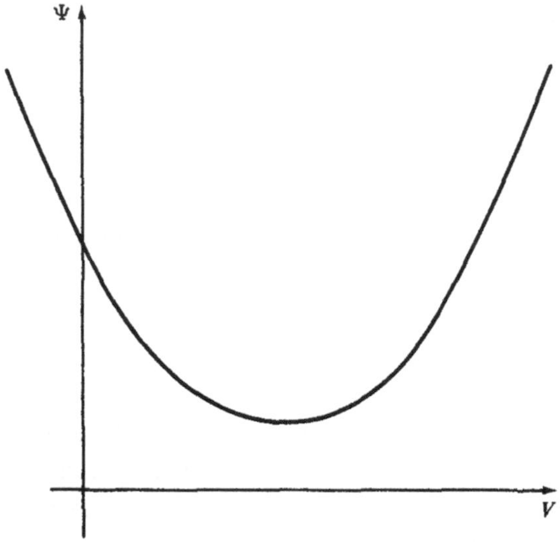

FIGURE 2.5. Helmholtz free energy $\Psi(V, T)$ as a function of V for fixed temperature T (convex upward function).

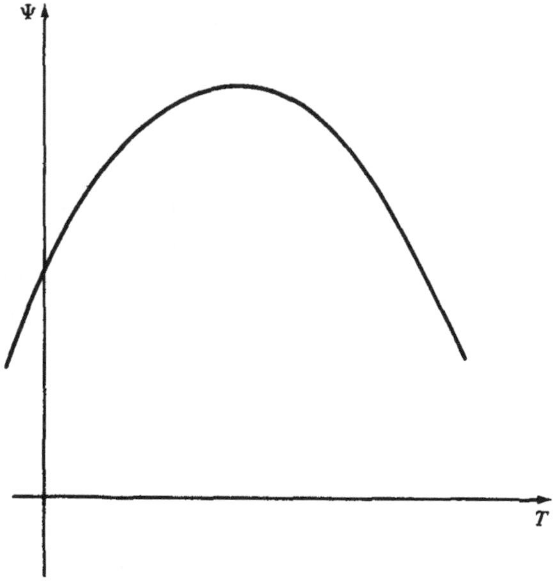

FIGURE 2.6. Helmholtz free energy $\Psi(V, T)$ as a function of T for fixed volume V (convex downward function).

with energy E a function of T only, i.e.,

$$E = \psi(T) \qquad \text{(Guy-Lussac's and/or Joule's law).} \tag{6.2}$$

For such systems the first law applied to an infinitesimal change gives

$$\begin{aligned}\delta Q &= P\,dV + dE \\ &= \frac{\phi(T)}{V}\,dV + \frac{d\psi}{dT}\,dT.\end{aligned} \tag{6.3}$$

Now from the second law $\delta Q/T$ is an exact differential; hence

$$\frac{\partial}{\partial T}\frac{\phi(T)}{TV} = \frac{\partial}{\partial V}\frac{1}{T}\frac{d\psi}{dT} = 0, \tag{6.4}$$

since $(1/T)(d\psi/dT)$ is a function only of T. It follows that

$$\frac{\phi(T)}{T} = \text{constant.} \tag{6.5}$$

The value R of this constant for 1 mole of gas,

$$R = 1.986 \text{ cal deg} \tag{6.6}$$

is called the *ideal-gas constant*. For 1 mole of ideal gas, therefore, the equation of state is given by

$$PV = RT, \tag{6.7}$$

and from the above argument this is the only system that can satisfy Boyle's law and Guy-Lussac's (or Joule's) law.

The internal energy E is given by

$$E = C_V T, \tag{6.8}$$

where C_V is the *specific heat* or *heat capacity* at constant volume defined by

$$C_V = \left(\frac{\delta Q}{dT}\right)_V = \left(\frac{\partial E}{\partial T}\right)_V, \tag{6.9}$$

where the second equality follows from the first law,

$$\delta Q = dU + P\,dV, \tag{6.10}$$

and $E(V, T) = U(P(V, T), V)$.

We define in a similar manner the specific heat at constant pressure C_P by

$$C_P = \left(\frac{\delta Q}{dT}\right)_P = P\left(\frac{\partial V}{\partial T}\right)_P + \left(\frac{\partial U}{\partial T}\right)_P \tag{6.11}$$

(considering U, or E, now as a function of P and T). For the ideal gas E is a function of T only; hence

$$\left(\frac{\partial E}{\partial T}\right)_P = \frac{dE}{dT} = C_V \tag{6.12}$$

and, from Equation 6.7,

$$P\left(\frac{\partial V}{\partial T}\right)_P = R. \tag{6.13}$$

It then follows from Equations 6.11, 6.12, and 6.13 that

$$C_P - C_V = R. \tag{6.14}$$

Equation 6.14 is a special case of a more general thermodynamic relation,

$$C_P - C_V = \frac{TV\alpha^2}{K_T}, \tag{6.15}$$

where

$$\alpha = \frac{1}{V}\left(\frac{\partial V}{\partial T}\right)_P \tag{6.16}$$

is the *coefficient of thermal expansion*, and K_T, defined by Equation 5.11, is the isothermal compressibility.

Since we shall refer back to this relation, we give a derivation of it here.

To prove Equation 6.15 we first consider V and T as the independent variables. From the first law we obtain

$$\begin{aligned} T\,dS &= dE + P\,dV \\ &= \left(\frac{\partial E}{\partial T}\right)_V dT + \left(\frac{\partial E}{\partial V}\right)_T dV + P\,dV \qquad (6.17) \\ &= C_V\,dT + \left[\left(\frac{\partial E}{\partial V}\right)_T + P\right] dV. \end{aligned}$$

Now since dS is an exact differential,

$$\frac{\partial}{\partial V}\left(\frac{C_V}{T}\right)_T = \left\{\frac{\partial}{\partial T}\left[\frac{1}{T}\left(\frac{\partial E}{\partial V}\right)_T + \frac{P}{T}\right]\right\}_V, \tag{6.18}$$

which immediately gives

$$\left(\frac{\partial E}{\partial V}\right)_T = T\left(\frac{\partial P}{\partial T}\right)_V - P. \tag{6.19}$$

Equation 6.17 then becomes

$$T\,dS = C_V\,dT + T\left(\frac{\partial P}{\partial T}\right)_V dV. \tag{6.20}$$

Similarly, if P and T are taken as independent variables we obtain

$$T\,dS = C_P\,dT - T\left(\frac{\partial V}{\partial T}\right)_P dP, \tag{6.21}$$

and if we subtract Equation 6.20 from 6.21 and use

$$dT = \left(\frac{\partial T}{\partial V}\right)_P dV + \left(\frac{\partial T}{\partial P}\right)_V dP, \tag{6.22}$$

which is obvious when P and V are taken as independent variables, we obtain

$$\begin{aligned}\left[(C_P - C_V)\left(\frac{\partial T}{\partial V}\right)_P - T\left(\frac{\partial P}{\partial T}\right)_V\right] dV \\ = -\left[(C_P - C_V)\left(\frac{\partial T}{\partial P}\right)_V - T\left(\frac{\partial V}{\partial T}\right)_P\right] dP.\end{aligned} \tag{6.23}$$

It then follows from Equation 6.23 and the chain relation (see Problem 5) connecting three dependent variables,

$$\left(\frac{\partial P}{\partial T}\right)_V\left(\frac{\partial T}{\partial V}\right)_P\left(\frac{\partial V}{\partial P}\right)_T = -1, \tag{6.24}$$

that

$$\begin{aligned}C_P - C_V &= T\frac{(\partial P/\partial T)_V}{(\partial T/\partial V)_P} \\ &= -T\left(\frac{\partial V}{\partial T}\right)_P^2\left(\frac{\partial P}{\partial V}\right)_T \\ &= \frac{TV\alpha^2}{K_T}\end{aligned} \tag{6.25}$$

where

$$\alpha = \frac{1}{V}\left(\frac{\partial V}{\partial T}\right)_P \quad \text{and} \quad K_T = -\frac{1}{V}\left(\frac{\partial V}{\partial P}\right)_T.$$

This is the required result, Equation 6.15.

After perusing the foregoing manipulation, the reader will conclude, almost correctly, that at this point thermodynamics becomes an exercise in implicit partial differentiation. This is then a good point to leave thermodynamics and move on to statistical mechanics.

PROBLEMS

1. Clausius' statement of the second law is the following: There is no transformation whose sole effect is to extract heat from a colder reservoir and transfer it to a hotter reservoir. Show that this statement is equivalent to Kelvin's statement. (*Hint:* Construct appropriate cycles to show that one statement is false when the other is false.)

2. By considering in turn the internal energy $U = U(V, S)$ as a function of volume and entropy, the Gibbs potential $\Phi = \Phi(P, T)$ as a function of pressure and temperature, and the *enthalpy* $H = U + PV$ as a function of pressure and entropy, deduce the following Maxwell relations:

$$T = (\partial U/\partial S)_V, \qquad P = -(\partial U/\partial V)_S;$$
$$S = -(\partial \Phi/\partial T)_P, \qquad V = (\partial \Phi/\partial P)_T;$$
$$V = (\partial H/\partial P)_S, \qquad T = (\partial H/\partial S)_P.$$

3. The specific heat at constant volume C_V is defined by

$$C_V = (\partial E/\partial T)_V,$$

where $E = E(V, T)$ is the internal energy. From the definition of the Helmholtz free energy $\Psi(V, T)$ deduce that $C_V \geq 0$ implies that $(\partial^2 \Psi/\partial T^2)_V \leq 0$, i.e., that the free energy is a concave function of temperature at fixed volume. What is the precise relationship between C_V and $(\partial^2 \Psi/\partial T^2)_V$?

4. We defined a convex function $f(x)$ in the text by $d^2f/dx^2 \geq 0$, assuming that $f(x)$ is second differentiable. This means that the function has the form shown in Figure 2.5. Prove that the following obvious geometrical facts are equivalent to the statement $d^2f/dx^2 \geq 0$ for second-differentiable functions.

(a) $f\left(\frac{x_1 + x_2}{2}\right) \leq \frac{1}{2}[f(x_1) + f(x_2)].$

That is, the midpoint of the curve between $f(x_1)$ and $f(x_2)$ lies below the midpoint of the chord joining $f(x_1)$ and $f(x_2)$.

(b) $f'(x_1) \leq \dfrac{f(x_2) - f(x_1)}{(x_2 - x_1)}$.

That is, tangents to the curve lie below the curve. [*Hint:* See Hardy et al. (B1964), p. 77.]

5. Let x, y, and z be three variables satisfying a functional relationship $f(x, y, z) = 0$. Deduce that

(a) $\left(\dfrac{\partial x}{\partial y}\right)_z = \left[\left(\dfrac{\partial y}{\partial x}\right)_z\right]^{-1}$,

(b) $\left(\dfrac{\partial x}{\partial y}\right)_z \left(\dfrac{\partial y}{\partial z}\right)_x \left(\dfrac{\partial z}{\partial x}\right)_y = -1$,

and if w is a function of any two of x, y, and z, that

(c) $\left(\dfrac{\partial x}{\partial y}\right)_w \left(\dfrac{\partial y}{\partial z}\right)_w = \left(\dfrac{\partial x}{\partial z}\right)_w$.

6. By considering adiabatic transformations ($dS = 0$), deduce from Equations 6.20 and 6.21 that

$$C_V = \frac{TV\alpha^2 K_S}{K_T(K_T - K_S)} \quad \text{and} \quad C_P = \frac{TV\alpha^2}{K_T - K_S},$$

where C_V and C_P are the specific heats at constant volume and pressure, respectively, $\alpha = V^{-1}(\partial V/\partial T)_P$ is the coefficient of thermal expansion,

$K_T = -V^{-1}(\partial V/\partial P)_T$ is the isothermal compressibility and,
$K_S = -V^{-1}(\partial V/\partial P)_S$ is the adiabatic compressibility.

CHAPTER 3

The Gibbs Ensembles and the Thermodynamic Limit

3-1 Introduction

The task of statistical mechanics is to derive macroscopic properties of matter from the laws governing the behavior of the individual particles. The macroscopic properties are expressed in terms of thermodynamic variables discussed in Chapter 2. In statistical mechanics one must first define the thermodynamic variables and then check that the laws of thermodynamics are valid. Unlike thermodynamics, however, which consists only of laws and relations between thermodynamic quantities (such as energy, specific heat, etc.), statistical mechanics aims to derive analytic expressions for these quantities.

Before we can proceed, we must specify the microscopic description. For simplicity we consider only classical systems with Hamiltonians of the form

$$H = \sum_{i=1}^{N} \left[\frac{\mathbf{p}_i^2}{2m} + U(\mathbf{r}_i)\right] + \sum_{1 \le i < j \le N} \phi(|\mathbf{r}_i - \mathbf{r}_j|), \tag{1.1}$$

where $\mathbf{p}_i^2/2m$ is the kinetic energy of the ith particle, $U(\mathbf{r}_i)$ is the potential at position $\mathbf{r}_i$ due to *outside* forces (e.g., walls) and we assume that the particles interact only pairwise through the central potential $\phi(r)$.

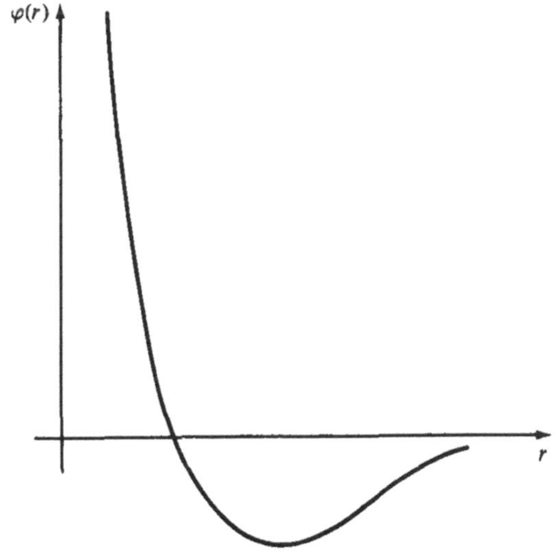

FIGURE 3.1. Form of the central pair potential φ (r), Equation 1.1.

Restrictions on $\phi(r)$ will be imposed later, but it is intuitively obvious that a "reasonable" potential must have the form shown in Figure 3.1. That is, it must be sufficiently repulsive at the origin to prevent "collapse" and vanish sufficiently rapidly at infinity to make the system stable (precise conditions are stated in Equations 6.14 and 6.15). Henceforth we shall consider only Hamiltonians of the type of Equation 1.1.

One could argue that quantum mechanics rather than classical mechanics is the more realistic microscopic description. This is no doubt true. The quantum-mechanical description, in fact, is conceptually simpler than the classical description, but the latter is somewhat easier to manipulate. In any case, classical statistical mechanics contains more than enough difficult and unsolved problems, so we shall only concern ourselves here with classical systems. Quantum statistical mechanics is discussed at length in the books by Huang (B1963) and Wannier (B1966), for example.

We shall now discuss the simplest type of system composed of N particles in a volume V with fixed energy E. The time evolution of the system is governed by Hamilton's equations (see Section 1-7)

$$\dot{p}_i = -\frac{\partial H}{\partial q_i}, \quad \dot{q}_i = \frac{\partial H}{\partial p_i}, \qquad i = 1, 2, \ldots, 3N, \tag{1.2}$$

where p_i and q_i are the canonical momenta and conjugate coordinates, respectively. Since we are considering only Hamiltonians that do not depend on time explicitly,

$$H(p, q) = E, \tag{1.3}$$

so the time evolution of a given state, specified by a point

$$(p, q) = (p_1, \ldots, p_{3N}, q_1, \ldots, q_{3N})$$

in $6N$-dimensional Γ space, is represented according to Equations 1.2 by a path in Γ space on the energy-surface equation (1.3).

We now discuss the equilibrium statistical mechanics of such a system.

3-2 The Gibbs Microcanonical Ensemble

On the macroscopic level we are interested in only a few quantities (e.g., the number of particles N, the volume V, and the energy E). We are not particularly interested in microscopic states of the system, since it is clear that a large number of different microscopic states will give the same macroscopic state. We consider, therefore, an ensemble of systems composed of the original system with all possible initial conditions. At time $t = 0$ we specify a density function $\rho(p, q, t = 0)$ on this ensemble which becomes a probability density when properly normalized. After time t, $\rho(p, q, t = 0)$ becomes $\rho(p, q, t)$, defined by

$$\int_A \rho(p, q, t)\, d^{3N}p\, d^{3N}q = \text{the number of points in the region } A \text{ of } \Gamma \text{ space at time } t. \tag{2.1}$$

The basic postulate due to Gibbs is the following.

POSTULATE I (GIBBS MICROCANONICAL ENSEMBLE). *The equilibrium distribution of the macroscopic states of an isolated mechanical system is the uniform distribution on the energy surface.*

In other words, the probability density $D_{\text{m.c.}}(p, q)$ in the microcanonical ensemble is given by

$$D_{\text{m.c.}}(p, q) = \frac{\rho(p, q)}{\int \rho(p, q)\, d\Gamma} = \frac{\delta(H - E)}{S(E)}. \tag{2.2}$$

$S(E)$ is the area of the energy surface $H = E$ (see Appendix A, Equation 16)

$$S(E) = \int \frac{d\sigma}{\|\text{grad } H\|}, \tag{2.3}$$

where $d\sigma$ is an element of energy surface and

$$\|\text{grad } H\|^2 = \sum_{i=1}^{3N} \left[\left(\frac{\partial H}{\partial p_i} \right)^2 + \left(\frac{\partial H}{\partial q_i} \right)^2 \right]. \tag{2.4}$$

$\delta(x)$ in Equation 2.2 is Dirac's delta function, so

$$\int \delta(H - E) \, d\Gamma = S(E). \tag{2.5}$$

(Hence $D_{\text{m.c.}}$ is properly normalized.)

Plausibility arguments can be given for this postulate and ergodic theorems can be used to make portions of the argument more precise. There is, however, only one physical system—hard spheres in a box—for which the ergodic theorems have been shown [Sinai (1963, 1966)] to be applicable. In any case, *there is no complete mathematical justification for the postulate, even if the ergodic theorems are applicable.* The basic problem is to reconcile macroscopic irreversibility with microscopic reversibility—in other words, to actually establish the existence of an equilibrium state.

We shall adopt the view that Equation 2.2 is the basic axiom of equilibrium statistical mechanics and proceed immediately to discuss its mathematical consequences. The interested reader is referred to Appendix B for a discussion of Gibbs' heuristic arguments and the relevance of ergodic theory.

3-3 Temperature and the Canonical Ensemble

Since we are not usually concerned in the real world with isolated systems, we turn our attention now to systems in contact with a heat reservoir. This enables us to introduce the notion of temperature for a system.

Consider a system S that is "weakly coupled" to, and in equilibrium with, a heat reservoir R. The combined system $S \cup R$ is taken to be isolated and described by the microcanonical distribution. We now state

POSTULATE II. *The conditional probability density for a system S in equilibrium with a system R is given by*

$$D = \frac{\exp(-\beta H_S)}{\int \exp(-\beta H_S) \, d\Gamma_S} \tag{3.1}$$

in the limit of an infinitely large heat reservoir R, where $\beta = (kT)^{-1}$, k is Boltzmann's constant, and T is the absolute temperature (of R), H_S is the Hamiltonian for the system S and $d\Gamma_S$ is an element of its phase space. The distribution (3.1) *is called the canonical distribution.*

Unlike the microcanonical distribution, a number of proofs can be given of Equation 3.1, granted the microcanonical postulate. All proofs, however, require a specific form for R, usually with a Hamiltonian of the form

$$H_R = H_1 + H_2 + \cdots + H_N, \tag{3.2}$$

where the H_i are weakly coupled to one another. Since such reservoirs are rarely, if ever, found in nature, it is clear that existing derivations of Equation 3.1 are not completely satisfactory from a physical point of view. Equation 3.1 may then be considered as an independent postulate. To give some (physical) justification for Equation 3.1, however, we now present a proof [due to Uhlenbeck and Ford (B1963)] for the special case where R is an ideal gas, i.e.,

$$H_R = \sum_{i=1}^{N} \frac{\mathbf{p}_i^2}{2m}. \tag{3.3}$$

Other proofs can be found in Khinchin (B1949) and Grad (1952).

We take the Hamiltonian for the combined system $S \cup R$ to be

$$H = H_S + H_R. \tag{3.4}$$

Strictly speaking, we should add an interaction term H_{RS} to Equation 3.4 (to make the combined system metrically transitive in the language of Appendix B), but since we can let H_{RS} vanish finally, we avoid this complication by setting $H_{RS} = 0$ in the beginning.

From Equation 2.2 the microcanonical distribution for H, Equation 3.4, is given by

$$D_{\text{m.c.}}\, d\Gamma_R\, d\Gamma_S = \frac{\delta(H_R + H_S - E)}{S(E)}\, d\Gamma_R\, d\Gamma_S, \tag{3.5}$$

where $S(E)$ is the area of the energy surface $H_R + H_S = E$ and $d\Gamma_R$, $d\Gamma_S$ are the phase-space elements for systems R and S, respectively.

The conditional distribution function D_S for system S is defined by

$$D_S\, d\Gamma_S = \left(\int D_{\text{m.c.}}\, d\Gamma_R\right) d\Gamma_S. \tag{3.6}$$

Now $d\Gamma_R = S_R(E_R)\, dE_R$, where $S_R(E)$ is the area of the reservoir energy surface, so, from Equation 3.5,

$$\begin{aligned} D_S &= [S(E)]^{-1} \int \delta(H_S + H_R - E) S_R(E_R)\, dE_R \\ &= [S(E)]^{-1} \int \delta(H_S + E_R - E) S_R(E_R)\, dE_R \\ &= \frac{S_R(E - H_S)}{S(E)}, \end{aligned} \tag{3.7}$$

where in the last step we have used the definition of the delta function. Normalization of D_S requires that

$$\begin{aligned} \int D_S\, d\Gamma_S &= [S(E)]^{-1} \int S_R(E - H_S) S_S(E_S)\, dE_S \\ &= [S(E)]^{-1} \int_0^E S_R(E - \varepsilon) S_S(\varepsilon)\, d\varepsilon = 1. \end{aligned} \tag{3.8}$$

That is,

$$S(E) = \int_0^E S_R(E - \varepsilon) S_S(\varepsilon)\, d\varepsilon. \tag{3.9}$$

[This can also be deduced from the definitions of $S(E)$, $S_R(E)$, and $S_S(E)$.]

Since we are taking R to be an ideal gas, enclosed, for example, in a cube of side L, the energy surface from Equation 3.3 is a hypercylinder whose base is a hypersphere of radius $(2mE)^{1/2}$ in $3N$ momentum space, and whose orthogonal cross section is a hypercube of side L in each of the $3N$ coordinate directions. It follows (see Problem 1) that

$$S_R(E) = C_N E^{(3N-1)/2}, \tag{3.10}$$

where we have indicated only the energy dependence. Equations 3.7 and 3.9 give

$$D_S = \frac{S_R(E - H_S)}{S_R(E)} \left[\int_0^E S_S(\varepsilon) \frac{S_R(E - \varepsilon)}{S_R(E)}\, d\varepsilon \right]^{-1}. \tag{3.11}$$

But, from Equation 3.10,

$$\begin{aligned} \lim_{N \to \infty} \frac{S_R(E - \alpha)}{S_R(E)} &= \lim_{N \to \infty} \left(1 - \frac{\alpha}{E}\right)^{(3N-1)/2} \\ &= \exp(-\beta\alpha), \end{aligned} \tag{3.12}$$

where

$$\beta = \frac{3N}{2E}. \tag{3.13}$$

Equation 3.11 in the limit ($N \to \infty$) of an infinite reservoir then becomes

$$D_S = \frac{\exp(-\beta H_S)}{\int_0^\infty S_S(\varepsilon)\exp(-\beta\varepsilon)\,d\varepsilon}. \tag{3.14}$$

That is

$$D_S = \frac{\exp(-\beta H_S)}{\int \exp(-\beta H_S)\,d\Gamma_S}, \tag{3.15}$$

which is the required result, Equation 3.1.

It is to be noted that β^{-1} is proportional to the energy per particle of the heat reservoir. In this sense it is independent of the "size" of R. Also, if we had several systems $S_1, S_2, \ldots, S_M$ coupled to and in equilibrium with R, we would arrive at the same distribution (3.15) for each S_i. β then has all the requirements of temperature and therefore must be a universal function of T. In fact, for an ideal gas $E = 3NkT/2$, where k is Boltzmann's constant and T the absolute temperature. We therefore take

$$\beta = (kT)^{-1} \tag{3.16}$$

in Equation 3.15.

One final point to note is that the argument leading to Equation 3.15 does not depend on the size of S. It depends only on the size of R. The canonical distribution is then valid for a one-particle system! We would, however, like the microcanonical and canonical ensembles to give equivalent thermodynamics. This means in particular that the energy in the canonical ensemble should be sharply peaked about its average value, which intuitively is the situation for large systems only. We shall return to this point later.

3-4 The Canonical Distribution and the Laws of Thermodynamics

Before we can proceed, we must set up a correspondence between statistical mechanical quantities and thermodynamic quantities and check that the laws of thermodynamics are satisfied.

Since we have presupposed equilibrium, the zeroth law is automatically satisfied. Also, since we are dealing at the outset with a conservative mechanical system, the total energy is conserved. The only problem with the *first law*, then, is the distinction between *quantity of heat* and *external work*.

To produce changes of state we assume that we have external fields that are characterized by a set of parameters $a_1, \ldots, a_k$. That is, $U(\mathbf{r}_i)$ in the Hamiltonian equation (1.1) is a function of $a_1, \ldots, a_k$. The work done on the gas for fixed $\mathbf{r}_i$'s by changing the a_i's is then

$$\delta W = -\sum_{i=1}^{k} \bar{X}_i \, \delta a_i, \tag{4.1}$$

where

$$\bar{X}_i = -\int d\Gamma D \frac{\partial H}{\partial a_i} \tag{4.2}$$

is the "force" in the direction of a_i averaged over all configurations of the gas. [D may be the microcanonical or the canonical distribution and it is assumed that $U(\mathbf{r}_i, a_1, \ldots, a_k)$ is a differentiable function of $a_1, a_2, \ldots, a_k$.]

We could take, for example,

$$U(\mathbf{r}_i, a_1, \ldots, a_k) = \sum_{j=1}^{k} \phi_j(|\mathbf{r}_i - \mathbf{R}_j|), \tag{4.3}$$

which describes a field produced by k fixed and independent external (or, in fact, internal) centers of force at positions $\mathbf{R}_i$. Also, if there is only one external field—the walls of the container—$a_1 = V$ and

$$\delta W = -P \, \delta V, \tag{4.4}$$

where, assuming that H is a differentiable function of V (i.e., the walls are represented by a smooth potential),

$$P = -\int d\Gamma D \frac{\partial H}{\partial V} \tag{4.5}$$

is the pressure exerted by the gas on the walls. (Note that if the walls are strictly hard walls, Equation 4.5 is not valid. The pressure in this case is then defined canonically, for example, by Equations 4.20, 4.22, and 4.23b.)

For an isolated system D is the microcanonical distribution $D_{\text{m.c.}}$ and it is obvious that $\delta W = \delta \bar{E}$ (i.e., δ is adiabatic), where the average energy $\bar{E}$ is defined by $\bar{E} = \int H D_{\text{m.c.}} \, d\Gamma$.

If, however, one takes the canonical distribution, $\bar{E}$, defined by

$$\bar{E} = \frac{\int H \exp(-\beta H) \, d\Gamma}{\int \exp(-\beta H) \, d\Gamma} \tag{4.6}$$

can be changed by changing β (i.e., the temperature) with the a_i's fixed. In this case it is obvious that $\delta W \neq \delta \bar{E}$ in general, so we *define*

$$\delta Q = \delta \bar{E} - \delta W \tag{4.7}$$

to be the quantity of heat put into the system. This takes care of the first law.

To show that the *second law* is satisfied we must show (see Section 2-4) from definitions 4.1, 4.6, and 4.7 that for a change δ in which both β and the a_i are changed reversibly (i.e., the system is always canonically distributed), $\beta\,\delta Q$ is an exact differential of the state of the system (i.e., a function of β and the a_i). (We shall not concern ourselves here with the irreversible part of the second law.)

To prove this, define the "partition function" Z by

$$Z = \int \exp(-\beta H)\, d\Gamma. \tag{4.8}$$

The average energy $\bar{E}$, Equation 4.6, is then given by

$$\bar{E} = -\frac{\partial}{\partial \beta} \log Z. \tag{4.9}$$

$\delta \bar{E}$ is given by

$$\delta \bar{E} = -\frac{\partial^2}{\partial \beta^2}(\log Z)\,\delta\beta - \sum_{i=1}^{k} \frac{\partial^2}{\partial \beta\, \partial a_i}(\log Z)\,\delta a_i \tag{4.10}$$

and δW defined by Equation 4.1 is

$$\begin{aligned}\delta W &= Z^{-1} \int \exp(-\beta H)\left(\sum_{i=1}^{k} \frac{\partial H}{\partial a_i}\,\delta a_i\right) d\Gamma \\ &= -\beta^{-1} \sum_{i=1}^{k} \frac{\partial}{\partial a_i}(\log Z)\,\delta a_i.\end{aligned} \tag{4.11}$$

We then obtain, from Equations 4.7, 4.10, and 4.11 (see Problem 2),

$$\begin{aligned}\beta\,\delta Q &= \beta(\delta \bar{E} - \delta W) \\ &= \delta\left[-\beta^2 \frac{\partial}{\partial \beta}(\beta^{-1} \log Z)\right],\end{aligned} \tag{4.12}$$

which implies the required result, that $\beta\,\delta Q$ is an exact differential of a state function.

Comparison with the thermodynamic formula (Equation 5.10 of Chapter 2) shows that the *entropy* S is given by

$$S = \frac{\partial}{\partial T}(kT \log Z) + \text{constant}, \tag{4.13}$$

where $\beta = (kT)^{-1}$, T is the absolute temperature, and k is a universal constant (Boltzmann's constant).

To determine k, consider an ideal gas enclosed in a volume V. The partition function Equation 4.8 is given by

$$\begin{aligned} Z &= \int \cdots \int_V d\mathbf{r}_1 \cdots d\mathbf{r}_N \int \cdots \int_{-\infty}^{\infty} d\mathbf{p}_1 \cdots d\mathbf{p}_N \exp\left(-\beta \sum_{i=1}^{N} \frac{\mathbf{p}_i^2}{2m}\right) \\ &= V^N (2\pi mkT)^{3N/2}. \end{aligned} \tag{4.14}$$

The pressure, Equation 4.5, is given by

$$\begin{aligned} P &= \frac{\partial}{\partial V}(kT \log Z) \\ &= \frac{NkT}{V}, \end{aligned} \tag{4.15}$$

which is the "ideal-gas law" if $k = R/N$ (see Equation 6.7 of Chapter 2) is the gas constant per particle.

It is clear that the additive constant in the entropy Equation 4.13 cannot be determined from the arguments presented here, so in a sense the constant is arbitrary. Its dependence on N, for example, can only be agreed upon by convention, and the normal convention requires that the entropy be an *extensive quantity*; i.e., for fixed T and for large N and V with fixed density $\rho = N/V$,

$$S(V, N, T) \sim Ns(\rho, T) \tag{4.16}$$

or, more precisely, the limit

$$\lim_{\substack{N, V \to \infty \\ \rho \text{ fixed}}} N^{-1} S(V, N, T) = s(\rho, T) \tag{4.17}$$

exists and depends only on the *intensive quantities* (i.e., independent of the size of the system) ρ and T.

In addition, one would like the classical results to agree in the $T \to \infty$ limit with the quantum-mechanical results if one makes the usual assumption that each nondegenerate energy level is "counted" only once (which is also a convention).

We shall not go into this here except to say that both objectives are realized if one defines the *canonical partition function* $Z(V, N, T)$ by (instead of Equation 4.8)

$$Z(V, N, T) = (N!\, h^{3N})^{-1} \int d\Gamma \exp(-\beta H), \tag{4.18}$$

where the Hamiltonian H will now be taken to be

$$H = \sum_{i=1}^{N} \frac{\mathbf{p}_i^2}{2m} + \sum_{1 \le i < j \le N} \phi(|\mathbf{r}_i - \mathbf{r}_j|) \tag{4.19}$$

with the understanding that the particles are confined to a volume V (i.e., the container has hard walls). The integral over momenta in Equation 4.18 is a trivial product of Gaussian integrals (see Equation 4.14), so

$$Z(V, N, T) = (N!\, \lambda^{3N})^{-1} \int \cdots \int_V d\mathbf{r}_1 \cdots d\mathbf{r}_N \exp\left[-\beta \sum_{i<j} \phi(|\mathbf{r}_i - \mathbf{r}_j|)\right], \tag{4.20}$$

where

$$\lambda = h(2\pi mkT)^{-1/2} \tag{4.21}$$

and h is Planck's constant.

At first sight it is rather disturbing to have a quantum mechanical constant (h) in a purely classical quantity (Z). Thermodynamic quantities (i.e., observables) defined in Equations 4.23 are, however, independent of h, as, of course, they should be.

Comparison with thermodynamic formulas (see Section 2-5) then shows that

$$\Psi(V, N, T) = -kT \log Z(V, N, T) \tag{4.22}$$

is the Helmholtz free energy, from which other thermodynamic quantities can be obtained by differentiation. For example, the entropy S, the pressure P, the internal energy $\bar{E}$, and the specific heat at constant volume C_V are given by

$$S = -\left(\frac{\partial \Psi}{\partial T}\right)_V, \tag{4.23a}$$

$$P = -\left(\frac{\partial \Psi}{\partial V}\right)_T, \tag{4.23b}$$

$$\bar{E} = \Psi - T\left(\frac{\partial \Psi}{\partial T}\right)_V, \tag{4.23c}$$

$$C_V = \left(\frac{\partial \bar{E}}{\partial T}\right)_V, \tag{4.23d}$$

and so forth.

Calculations then in the canonical ensemble essentially begin and end with an evaluation of the partition function Equation 4.20.

We will show in Section 3-6 that *for suitable pair potentials* $\phi(r)$ the limit

$$\lim_{\substack{N, V \to \infty \\ \rho = N/V \text{ fixed}}} N^{-1}\Psi(V, N, T) = \psi(\rho, T) \tag{4.24}$$

exists and is a function only of density ρ and temperature T. The *extensive* property, Equation 4.17, for the entropy follows from 4.24 and 4.23a. Whenever the limit in Equation 4.24 exists, the thermodynamics is called extensive. The limiting process is called the *thermodynamic limit*, which is the subject of Section 3-6.

3-5 The Grand-canonical Distribution

Before discussing the existence of the thermodynamic limit and other matters, we briefly mention one last distribution, which in many ways is more useful than either the canonical or microcanonical distributions.

Recall that the microcanonical distribution applies to an isolated system with fixed N, V, and E, and that the canonical distribution applies to a system with fixed N, V, and T and variable energy E. The next obvious step is the grand-canonical distribution, which applies to a system in which T and V are fixed and both N and E are allowed to vary.

Mathematically, the grand-canonical distribution is simply the generating function for the canonical partition function $Z(V, N, T)$, Equation 4.20, i.e.,

$$Z_G(V, T, z) = \sum_{K=0}^{\infty} Z(V, K, T)z^K, \tag{5.1}$$

where the variable z is called the *fugacity* or activity.

The pressure P (as a function of V, T, and z) in the grand-canonical ensemble is defined by

$$PV = kT \log Z_G(V, T, z) \tag{5.2}$$

and the (average) number of particles is given by

$$N = z \frac{\partial}{\partial z} \log Z_G(V, T, z). \tag{5.3}$$

The equation of state is obtained from Equations 5.2 and 5.3 by eliminating the fugacity z.

In the grand-canonical ensemble the internal energy is given by (see Equation 4.23c)

$$E = -\frac{\partial}{\partial \beta} \log Z_G(V, T, z), \tag{5.4}$$

the Helmholtz free energy $\Psi(V, T, z)$ is given by

$$\Psi(V, T, z) = NkT \log z - kT \log Z_G(V, T, z), \tag{5.5}$$

and so on, where it is to be understood that z is now a function of N, V, and T (from Equation 5.3).

It is now natural to ask whether the different ensembles—microcanonical, canonical, and grand canonical—give the same thermodynamics.

Consider, for example, the microcanonical and the canonical. The canonical ensemble contains systems of all energies, so to show that the two ensembles are equivalent, we must show that the energy distribution is sharply peaked about the (canonical) average value.

The mean-square fluctuation of energy in the canonical ensemble is given by (see Problem 4)

$$\begin{aligned} \langle (H - \langle H \rangle)^2 \rangle &= \langle H^2 \rangle - \langle H \rangle^2 \\ &= kT^2 C_V, \end{aligned} \tag{5.6}$$

where

$$\langle H \rangle = \frac{\int H \exp(-\beta H)\, d\Gamma}{\int \exp(-\beta H)\, d\Gamma} \tag{5.7}$$

is the average energy, C_V is the specific heat at constant volume defined by Equation 4.23d, etc.

For large N we expect $\langle H \rangle$ and C_V to be extensive (i.e., proportional to N), so, from Equation 5.6, the root-mean-square deviation of the energy from its mean value

$$\left[\frac{\langle (H - \langle H \rangle)^2 \rangle}{\langle H \rangle^2}\right]^{1/2} = \left[\frac{kT^2 C_V}{\langle H \rangle^2}\right]^{1/2} \tag{5.8}$$

would be of order $N^{-1/2}$. Thus in the thermodynamic limit, $N, V \to \infty$ with fixed density $\rho = N/V$, the two ensembles are equivalent provided the system has the extensive property (i.e., if the thermodynamic limit exists). A similar conclusion holds for the grand-canonical and canonical ensembles.

It is obvious that the different ensembles will not give identical thermodynamics for finite systems, so from this point of view the limit is essential. We will see in Chapter 4 that it is also essential for a rigorous discussion of phase transitions. The question then arises: For what (pair) potentials does the thermodynamic limit exist?

3-6 Existence of the Thermodynamic Limit

We shall formulate the problem for the canonical ensemble only and prove the existence for a special class of potentials. The known existence theorems for the classical and quantum-mechanical microcanonical, canonical, and grand-canonical ensembles will be stated at the end of this section.

First we state more precisely what we mean by the thermodynamic limit (T.L. for short): $N \to \infty$, $V \to \infty$ with fixed density $\rho = N/V$. Consider a sequence of domains $\{\Omega_k : k = 1, 2, \ldots\}$ with volumes $V_k < V_{k+1}$ containing N_k particles such that N_k/V_k is fixed (more generally we could take $N_k \to \infty$, $V_k \to \infty$ as $k \to \infty$ such that $N_k/V_k \to \rho$ as $k \to \infty$). Define the partition function for the kth domain to be

$$Z_k(\Omega_k, N_k, T) = (N_k!\,\lambda^{3N_k})^{-1} \int_{\Omega_k}\!\cdots\!\int d\mathbf{r}_1 \cdots d\mathbf{r}_{N_k} \exp\Big[-\beta \sum_{i<j} \phi(|\mathbf{r}_i - \mathbf{r}_j|)\Big] \tag{6.1}$$

and the corresponding free energy per particle ψ_k by

$$f_k = -\beta\psi_k = N_k^{-1} \log Z_k \tag{6.2}$$

(we shall refer to f_k also as the free energy per particle). The problem is: For what potentials $\phi(r)$ and for what domains Ω_k does the T.L.

$$\lim_{k\to\infty} f_k = f(\rho, T) \tag{6.3}$$

exist?

We shall not dwell on the domain question except to say that almost any shapes will do except those for which the surface area increases too rapidly (more than $V^{2/3}$ is too fast). The potential question is clearly of more interest.

We shall consider the special case where the potential $\phi(r)$ has the form shown in Figure 3.2, i.e.,

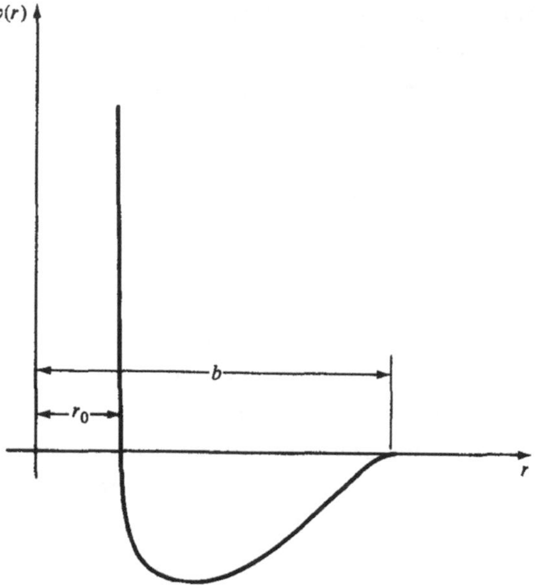

FIGURE 3.2. Van Hove potential (Equation 6.4).

$$\phi(r) = \begin{cases} \infty & r \leq r_0, \\ <0 & r_0 < r < b, \\ 0 & r \geq b, \\ > -\varepsilon & r > r_0. \end{cases} \tag{6.4}$$

In other words, the potential has a *hard core* of radius r_0, and a *finite-range* (*b*) *attractive tail*. Notice that the "hard core" simply means that the domain of integration in Equation 6.1 is restricted to $|\mathbf{r}_i - \mathbf{r}_j| > r_0$ for all $i, j = 1, \ldots, N$. We call this potential the *Van Hove potential*, after Leon Van Hove, who was the first [Van Hove (1949)] to prove that the limit (Equation 6.3) exists for this potential. Unfortunately, Van Hove's proof is incomplete. The argument, however, has since been tightened up and generalized by Ruelle (1963), Fisher (1964), and others. The proof given here is due to Ruelle (1963) [see also Ruelle (B1969)].

Before we can proceed we must specify the sequence of domains $\{\Omega_k : k = 1, 2, \ldots\}$. For simplicity we take a *standard sequence of cubes* defined as follows. Start with a cube Ω_1, with "free volume" V_1, and walls of thickness

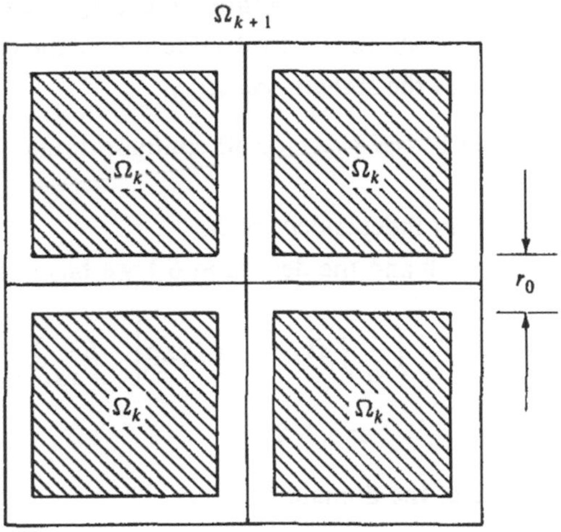

FIGURE 3.3. Standard sequence of cubes; the free volume is shaded.

$r_0/2$. Proceeding inductively, we construct domain Ω_{k+1} by placing eight Ω_k cubes with free volumes V_k and walls of thickness $r_0/2$, in a larger cube with free volume $V_{k+1} = 8V_k$ and walls of thickness $r_0/2$, as shown in Figure 3.3. The number of particles in domain Ω_{k+1} is $N_{k+1} = 8N_k$ and the partition function for the kth domain is defined by Equation 6.1.

Since the interactions between the eight Ω_k cubes is purely attractive (from Equation 6.4), i.e., $\phi(|\mathbf{r}_1 - \mathbf{r}_2|) \leq 0$ when $\mathbf{r}_1$ and $\mathbf{r}_2$ are in different cubes, we decrease the integrand in

$$Z_{k+1}(\Omega_{k+1}, N_{k+1}, T) = (N_{k+1}!\,\lambda^{3N_{k+1}})^{-1} \int_{\Omega_{k+1}} \cdots \int \exp\left[-\beta \sum_{1 \leq i < j \leq N_{k+1}} \phi(|\mathbf{r}_i - \mathbf{r}_j|)\right] d\mathbf{r}_1 \cdots d\mathbf{r}_{N_{k+1}} \tag{6.5}$$

by eliminating interactions between particles in different Ω_k cubes. The domain of integration is also decreased by restricting N_k of the $N_{k+1} = 8N_k$ particles to be in each of the cubes Ω_k. Since there are $(8N_k)!/(N_k!)^8$ ways of arranging $8N_k$ particles in eight cubes with N_k in each, it is obvious from Equation 6.5 that

$$Z_{k+1}(\Omega_{k+1}, N_{k+1}, T) > [Z_k(\Omega_k, N_k, T)]^8. \tag{6.6}$$

From the definition of f_k, Equation 6.2, we then obtain

$$f_{k+1} > f_k. \tag{6.7}$$

The sequence $\{f_k: k = 1, 2, \ldots\}$ is therefore increasing and will have a limit if the sequence is bounded above. This will certainly be the case if, in general, the potential $\phi(r)$ satisfies the *stability condition*

$$\sum_{1 \le i < j \le N} \phi(|\mathbf{r}_i - \mathbf{r}_j|) > -NB \qquad \text{for all configurations } \mathbf{r}_1, \ldots, \mathbf{r}_N, \tag{6.8}$$

where $B > 0$ is a constant independent of N.

Thus from Equation 6.8 and the definition 6.1 we have that

$$Z(\Omega, N, T) = (N!\,\lambda^{3N})^{-1} \int \cdots \int_\Omega d\mathbf{r}_1 \cdots d\mathbf{r}_N \exp\left[-\beta \sum_{1 \le i < j \le N} \phi(|\mathbf{r}_i - \mathbf{r}_j|)\right]$$
$$< (N!\,\lambda^{3N})^{-1} V^N \exp(NB) \tag{6.9}$$

where V is the volume of the domain Ω. Since

$$\log N! > N \log N - N \tag{6.10}$$

it follows from Equation 6.9 that

$$\begin{aligned} f &= N^{-1} \log Z(\Omega, N, T) \\ &< \beta B + 1 - 3 \log \lambda - \log\left(\frac{N}{V}\right), \end{aligned} \tag{6.11}$$

and since we are keeping N/V fixed, Equation 6.11 gives the required upper bound.

Our proof of the existence of the T.L. will then be complete once we have shown that the potential (Equation 6.4) is *stable* (i.e., satisfies Equation 6.8). This is immediate since from $\phi(r) = 0$ for $r \ge b$, it follows that only a finite number of particles, the number α of spheres of radius r_0 which can be packed into a sphere of radius b, can interact with a given particle. In other words, since $\phi(r) > -\varepsilon$,

$$\sum_{j=1}^{N} \phi(|\mathbf{r}_i - \mathbf{r}_j|) > -\alpha\varepsilon \qquad \text{for all } \mathbf{r}_i, \mathbf{r}_j, \qquad i, j = 1, 2, \ldots, N. \tag{6.12}$$

Thus

$$\sum_{1 \le i < j \le N} \phi(|\mathbf{r}_i - \mathbf{r}_j|) > -N\alpha\varepsilon \quad \text{for all configurations } \mathbf{r}_1, \ldots, \mathbf{r}_N, \tag{6.13}$$

and the proof is complete.

Notice that we could arrive at the conclusion (Equation 6.6) by relaxing the condition $\phi(r) = 0$ for $r \ge b$ and replacing it by the *weak tempering condition* $\phi(r) < 0$ for $r > R$. The above proof (to Equation 6.7) can then be repeated word for word if the cubes Ω_k are taken to have walls of thickness $R/2$ instead of $r_0/2$. More general domains than the standard sequence of cubes can be considered if one "fills" domains with cubes a certain distance apart. The inequality 6.7 then contains "surface terms" which go to zero in the limit for "reasonable domains," and the proof goes through, essentially as above, without too much difficulty. The interested reader is referred to Fisher (1964) and Ruelle (B1969) for details.

Generally speaking, we need two conditions on the potential, one for large distances to give the inequality 6.7 and one at small distances to give the inequality 6.8. Fisher has shown that sufficient conditions for the T.L. to exist are

$$\begin{aligned}&\text{(a)}\quad \phi(r) \geq -a \\ &\text{(b)}\quad |\phi(r)| \leq \frac{C}{r^{d+\varepsilon}} \qquad \text{as } r \to \infty\end{aligned} \tag{6.14}$$

and

$$\text{(c)}\quad \phi(r) \geq \frac{C'}{r^{d+\varepsilon'}} \qquad \text{as } r \to 0, \tag{6.15}$$

where d is the dimensionality of the system and C, C', ε, and ε' are arbitrary positive numbers. Ruelle and Fisher have shown, moreover, that under these conditions the T.L. for the grand-canonical ensemble exists and gives the same thermodynamics as the canonical ensemble. Griffiths (1965b) has discussed the equivalence of the microcanonical and canonical ensembles, and all results carry over to quantum-mechanical systems.

Until very recently a long-outstanding problem was the question of the existence of the T.L. for systems with pure Coulomb forces (without hard cores), i.e.,

$$\phi(|\mathbf{r}_i - \mathbf{r}_j|) = \frac{e_i e_j}{|\mathbf{r}_i - \mathbf{r}_j|} \tag{6.16}$$

where e_i is the charge of the ith particle. Thanks to the combined efforts of Dyson, Lenard, Lebowitz, and Lieb, this problem has now been solved.

It is almost obvious first of all that a classical system with interaction 6.16 will be unstable (i.e., the inequality 6.8 will not hold), so quantum mechanics is essential. Dyson and Lenard (1967) and Lenard and Dyson (1968) showed that one has "quantum-mechanical stability," i.e.,

$$E_0 > -NB \tag{6.17}$$

where E_0 is the ground-state energy of the system, if and only if at least one of the species of particles obeys Fermi statistics (i.e., the exclusion principle), which plays, if you like, the role of a "hard core." The condition 6.17 ensures that the free energy f is bounded above, but the proof given recently by Lebowitz and Lieb (1969) that the limit actually exists requires a more judicious choice of domains than cubes. What one takes essentially, instead of a standard sequence of cubes, is an infinite sequence of smaller and smaller spheres which gives, instead of Equation 6.7, assuming charge neutrality, a "renewal inequality,"

$$f_k \geq a_1 f_{k-1} + a_2 f_{k-2} + \cdots + a_k f_0 \qquad \left(\sum_{j=1}^{\infty} a_j = 1\right). \tag{6.18}$$

The existence of the limit can then be deduced from the Dyson–Lenard result and some results of renewal theory.

The reason for choosing spheres is I. Newton's classic observation that two nonoverlapping spheres with spherically symmetric charge distribution interact as though their total charges are at their centers. In the construction, then, one simply maintains charge neutrality in each of the smaller spheres.

3-7 Convexity and Thermodynamic Stability

To show that thermodynamics exists for a given system, it is not enough to prove the extensive property (i.e., the existence of the thermodynamic limit). One must also show that the resulting thermodynamics is stable. That is, the compressibility and specific heat, for example, are nonnegative (see the discussion in Section 2-5). These results essentially follow from convexity properties of the free energy, which in turn follow almost immediately from the existence proof given in the previous section (and generalizations of it).

Thus for the Van Hove potential (Equation 6.4) one constructs Ω_{k+1} from Ω_k as before but now one places N_k^1 particles in four of the Ω_k cubes and N_k^2 particles in the remaining four Ω_k cubes. Keeping the densities $\rho_1 = N_k^1/V_k$ and $\rho_2 = N_k^2/V_k$ fixed, we obtain, instead of Equations 6.6 and 6.7,

$$Z_{k+1}(\Omega_{k+1}, 4N_k^1 + 4N_k^2, T) \geq [Z_k(\Omega_k, N_k^1, T)]^4 [Z_k(\Omega_k, N_k^2, T)]^4 \tag{7.1}$$

and

$$g_{k+1}\left(\frac{\rho_1+\rho_2}{2}\right) \geq \tfrac{1}{2}[g_k(\rho_1)+g_k(\rho_2)], \tag{7.2}$$

respectively, where $g_k(\rho_i)$ is the "free energy per unit volume" defined by

$$g_k(\rho_i) = V_k^{-1} \log Z_k(\Omega_k, N_k^i, T), \qquad i = 1, 2. \tag{7.3}$$

Notice that g_k is related to f_k defined previously (Equation 6.2) by

$$g_k(\rho) = \frac{N_k}{V_k} f_k = \rho f_k \tag{7.4}$$

and since $\lim_{k\to\infty} f_k = f$ exists, $\lim_{k\to\infty} g_k = g$ also exists. Taking the limit $k \to \infty$, Equation 7.2 becomes

$$g\left(\frac{\rho_1+\rho_2}{2}\right) \geq \tfrac{1}{2}[g(\rho_1)+g(\rho_2)], \tag{7.5}$$

i.e., *$g(\rho)$ is a concave (or convex downward) function of the density ρ, as shown in Figure* 3.4. From Equation 7.4,

$$g(\rho) = v^{-1} f(v), \tag{7.6}$$

where $v = \rho^{-1}$ is the "specific volume."

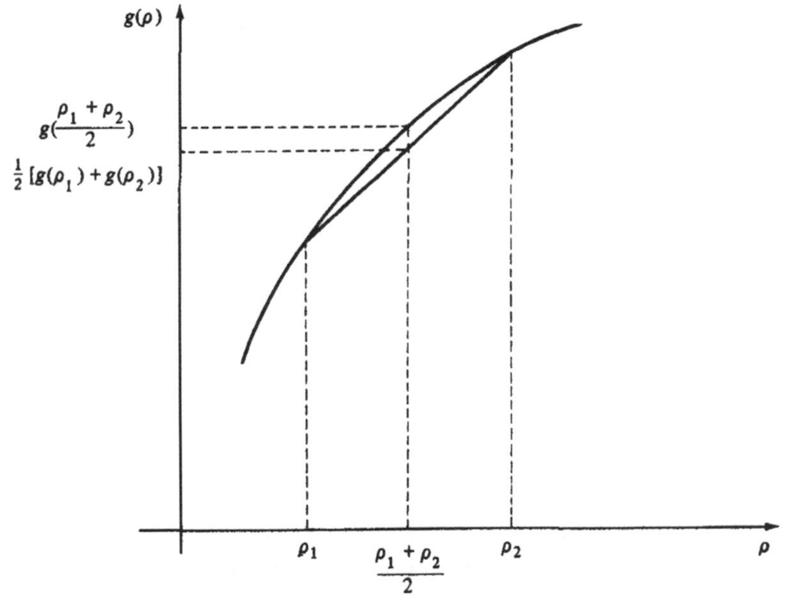

FIGURE 3.4. Convex downward function $g(\rho)$. The chord joining two points always lies below the curve.

The following properties of convex functions (i.e., convex downward functions) are now important [see Hardy *et al.* (B1964)]:

1. A bounded convex function is continuous.
2. Equation 7.5 implies complete convexity, i.e.,

$$g\left(\sum_{i=1}^{n} \omega_i \rho_i\right) \geq \sum_{i=1}^{n} \omega_i g(\rho_i) \tag{7.7}$$

for all nonnegative ω_i such that $\sum_{i=1}^{n} \omega_i = 1$.

3. $dg/d\rho$ exists almost everywhere and left and right derivatives exist everywhere. The right-hand derivative, moreover, is not greater than the left-hand derivative, and both derivatives are nonincreasing functions of ρ.
4. Convexity defined by Equation 7.5 has the geometrical interpretation depicted in Figure 3.4; i.e., the value of the function at the midpoint of an arc is not less than the mean value of the function at the two end points of the arc. Most of the properties of convex functions can be seen directly from the figure (see Problem 4 of Chapter 2).

It follows from property 1, since g is bounded (Equations 6.11 and 7.4), that $g(\rho)$ is a continuous function of ρ, and hence that the free energy $f(v)$ (Equation 7.6) is a continuous function of v.

Property 2 implies that $f(v)$ is a convex function of v (see Problem 7). Also, from Equation 7.7 with $n = 2$, $\rho_1 = 0$, $\rho_2 = \rho$, and $\omega_2 = \omega$

$$g(\omega\rho) \geq \omega g(\rho) \qquad \text{when } 0 \leq \omega \leq 1 \tag{7.8}$$

since from Equation 7.6 and the definition of $f(v)$, $g(0) = 0$. Equations 7.6 and 7.8 imply that

$$f(v) \leq f(v') \qquad \text{when } v \leq v' \tag{7.9}$$

i.e., that $f(v)$ is a monotonic nondecreasing function of v.

By definition the pressure (in the T.L.) is given by

$$p = kT\frac{\partial f}{\partial v}, \tag{7.10}$$

which exists almost everywhere (property 3) and, from Equation 7.9, is nonnegative. Also, since the derivative of a convex function is necessarily nonincreasing, the *pressure is a monotonic nonincreasing function of the specific volume v*. Since $p(v)$ is monotonic, its derivative exists almost everywhere and hence the inverse of the isothermal compressibility

$$K_T^{-1} = -v\frac{dp}{dv} \tag{7.11}$$

exists and is nonnegative almost everywhere.

To prove that the pressure is actually a continuous function of v requires a little more effort, since the derivative of a convex function may have jump discontinuities. Ruelle first proved absolute continuity for classical systems with the additional assumption that $|\phi(r)|$ is bounded. Ruelle's result has recently been extended by Ginibre (1967) and others to potentials of essentially Van Hove type [see Ruelle (B1969)].

Finally, it is a relatively easy matter to show that $f(v, T)$ is a convex (upward) function of T, from which it follows that the specific heat is nonnegative (see Problem 4).

Similar results hold for the other ensembles and for quantum-mechanical systems. The interested reader is referred to Ruelle (B1969) for details.

PROBLEMS

1. Show that

$$\underset{\sum_{i=1}^{n} x_i^2 = a^2}{\int \cdots \int} d\sigma_a = \frac{2\pi^{n/2}a^{n-1}}{\Gamma(n/2)},$$

where $d\sigma_a$ is the surface element of the n-dimensional sphere with radius a and $\Gamma(\alpha)$ is the gamma function. [*Hint:* Use the fact that

$$\left(\frac{\pi}{\lambda}\right)^{n/2} = \int_{-\infty}^{\infty}\cdots\int \exp\left(-\lambda\sum_{i=1}^{n} x_i^2\right) dx_1 \cdots dx_n$$

$$= \int_0^{\infty} ds \exp(-\lambda s^2) \underset{\sum_{i=1}^{n} x_i^2 = s^2}{\int \cdots \int} d\sigma_s$$

and after a suitable change of variables consult tables of Laplace transforms.]

2. If

$$\delta E = -\frac{\partial^2}{\partial\beta^2}(\log Z)\,\delta\beta - \sum_k \frac{\partial^2}{\partial\beta\,\partial a_k}(\log Z)\,\delta a_k$$

and

$$\delta W = -\beta^{-1} \sum_k \frac{\partial}{\partial a_k} (\log Z)\, \delta a_k$$

(Z is the canonical partition function) are the changes in internal energy and work done on a system resulting from changes in temperature and external field parameters a_k, show that $\beta(\delta \bar{E} - \delta W)$ is an exact differential of the function

$$-\beta^2 \frac{\partial}{\partial \beta} (\beta^{-1} \log Z) + \text{constant}.$$

3. (a) Calculate the pressure of a classical ideal gas (Hamiltonian $H = \sum_{i=1}^{n} \mathbf{p}_i^2/2m$) as a function N, V, and T by calculating the canonical partition function.

(b) Do the same using the grand-canonical partition function.

4. If Z is the partition function and an average $\langle A \rangle$ of a function A of the state of the system is defined by

$$\langle A \rangle = Z^{-1} \int d\Gamma\, A \exp(-\beta H),$$

where H is the Hamiltonian of the system, show that

$$\langle H^2 \rangle - \langle H \rangle^2 = kT^2 C_V$$

and deduce from this that the specific heat at constant volume C_V is non-negative.

5. From the equations

$$\Psi = NkT \log z - kT \log Z_G(z, V, T)$$

and

$$N = z \frac{\partial}{\partial z} \log Z_G(z, V, T)$$

for the grand-canonical ensemble, show that the *chemical potential* μ satisfies

$$\mu = \left(\frac{\partial \Psi}{\partial N}\right)_{V,T} = kT \log z.$$

6. (a) Show that in the grand-canonical ensemble the mean-square fluctuation of the number of particles can be expressed in the form

$$\langle N^2\rangle - \langle N\rangle^2 = z\frac{\partial}{\partial z} z \frac{\partial}{\partial z} \log Z_G(z, V, T).$$

(b) Show that the statement in part (a) is the same as

$$\langle N^2\rangle - \langle N\rangle^2 = -\frac{kTN}{\left(V^2 \dfrac{\partial P}{\partial V}\right)}$$

by first noting that the right-hand side is

$$\left(\frac{\partial}{\partial N}\log z\right)^{-1}$$

and then using the result of Problem 5.

7. If $f(v) = vg(1/v)$, $v > 0$, and $g(\rho)$ is a continuous convex function of ρ, i.e.,

$$g\left(\sum_{i=1}^{n} \omega_i \rho_i\right) \geq \sum_{i=1}^{n} \omega_i g(\rho_i),$$

where $\omega_i \geq 0$ and $\sum_{i=1}^{n} \omega_i = 1$, deduce that $f(v)$ is also a convex function of v, i.e., that

$$f\left(\frac{v_1 + v_2}{2}\right) \geq \tfrac{1}{2}[f(v_1) + f(v_2)].$$

8. If the potential between two particles at positions $\mathbf{r}_i$ and $\mathbf{r}_j$ satisfies

$$\phi(|\mathbf{r}_i - \mathbf{r}_j|) \leq \frac{C}{|\mathbf{r}_i - \mathbf{r}_j|^{d+\varepsilon}} \qquad \text{for } |\mathbf{r}_i - \mathbf{r}_j| \geq R$$

(d is dimensionality and C and ε are positive), show that for any two subdomains Ω_1 and Ω_2 of a domain Ω such that the minimum distance between Ω_1 and Ω_2 is at least R,

$$Z(\Omega, N_1 + N_2, T) \geq Z(\Omega_1, N_1, T)Z(\Omega_2, N_2, T)\exp\left(-\frac{N_1 N_2 \beta C}{R^{d+\varepsilon}}\right).$$

CHAPTER 4

Phase Transitions and Critical Points

4-1 Introduction

One of the most interesting and challenging problems in equilibrium statistical mechanics is the problem of phase transitions. We are all familiar with boiling water and the transition of a gas to a liquid under compression, which are simple examples of phase transitions. Mathematically, the problem is to explain, or derive, the existence of phase transitions and the behavior of thermodynamic quantities in the neighborhood of the transition point from the statistical-mechanical ensembles.

Thermodynamically we describe, for example, the liquid–gas transition as follows.

Consider a gas at a fixed temperature T. If the gas is compressed at sufficiently small T it reaches a certain density ρ_G (or specific volume $v_G = \rho_G^{-1}$) and then condenses at constant pressure to a liquid of greater density ρ_L (or smaller specific volume $v_L = \rho_L^{-1}$). This is represented diagrammatically in the p–v diagram shown in Figure 4.1.

The full curves are the *isotherms* (p as a function of v for fixed temperature T) and the dashed curve is the coexistence curve (i.e., inside the dashed curve gas and liquid coexist).

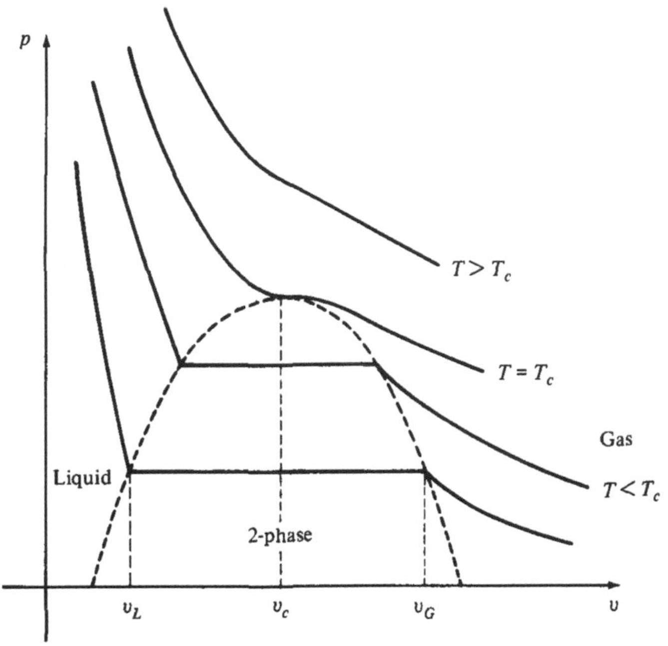

FIGURE 4.1. Isotherms for a gas–liquid phase transition.

At a certain temperature called the critical temperature, T_c, the gas remains a gas no matter how much it is compressed. The change occurs abruptly and is usually characterized by singular behavior in the thermodynamic functions (e.g., along the *critical isochore* $v = v_c$ the specific heat and the compressibility diverge at $T = T_c$).

Mathematically, then, a phase-transition point is a singular point of the partition function

$$Z(V, N, T) = (N!\,\lambda^{3N})^{-1} \int_V \cdots \int d\mathbf{r}_1 \cdots d\mathbf{r}_N \exp\left[-\beta \sum_{1 \le i < j \le N} \phi(|\mathbf{r}_i - \mathbf{r}_j|)\right]. \tag{1.1}$$

It is immediately obvious, however, that for finite N, V, and T, $Z(V, N, T)$ is a completely analytic function of T [the integrand is an analytic function of $\beta = (kT)^{-1}$ and the integral is over a finite domain]. Hence, to have a phase transition in the mathematical sense, it is clearly necessary to first take the thermodynamic limit, $N, V \to \infty$ with fixed density $\rho = N/V$. After this has been done, we define a phase-transition point precisely as follows.

Any nonanalytic point of the canonical free energy

$$f(v, T) = \lim_{\substack{N, V\to\infty \\ v=V/N \text{ fixed}}} N^{-1} \log Z(V, N, T) \tag{1.2}$$

or the grand-canonical potential

$$\chi(z, T) = \lim_{V\to\infty} V^{-1} \log Z_G(V, z, T) \tag{1.3}$$

occurring for real positive T, v, or z is called a phase-transition point.

By nonanalytic we mean that the function cannot be expanded in a Taylor series about the point in question. It is not necessary, therefore, for any particular derivative of $f(v, T)$ or $\chi(z, T)$ to have a discontinuity or become infinite, but normally this is what happens.

For a long time (until about 1940, in fact) it was thought that points so defined would not exist, the traditional view being: How are molecules to know when they should condense? It was felt that extra conditions would be needed to tell the system which phase to go into. But thanks to Onsager, who in 1944 showed by an example that we shall discuss in Chapter 5 that a phase transition can result from the partition function alone, it is now commonly felt that the partition function contains all the information necessary for a phase transition. One of the outstanding problems to this day, though, is to find the essential information necessary to characterize a phase transition.

It is obvious to begin with that an evaluation of the partition function 1.1 in general is a hopelessly difficult task, so most of the effort to date has been spent on very special cases. Even for perhaps the simplest case of a purely hard-core potential (i.e., the particles are hard noninteracting spheres),

$$\phi(r) = \begin{cases} \infty & r \le r_0 \\ 0 & r > r_0 \end{cases} \tag{1.4}$$

it is not known if there is a phase transition for dimensions higher than one.

In one dimension the hard-core potential corresponds to a gas of hard rods. This problem was solved many years ago by Tonks (1936). Some extensions of the Tonks gas can also be solved and we discuss one of these, the Takahashi gas, along with the Tonks gas in the next section. In the present context they are somewhat uninteresting since they do not have phase transitions. In fact, Van Hove (1950) proved that a one-dimensional gas of hard rods with a finite-range attraction cannot have a phase transition. There has been considerable interest recently in finding under what conditions (on the potential) one can have a phase transition in one dimension. Ruelle (and co-workers) [see Ruelle (B1969)] and Dyson (1969a,b) have recently proved some theorems in this direction which we mention briefly at the end of the following section.

4-2 The One-dimensional Tonks and Takahashi Gases

Consider a collection of N particles in the interval $(0, L)$ interacting pairwise through a potential $\phi(x)$. The problem is to evaluate the partition function defined by

$$Z(L, N, T) = (N!)^{-1} \int_0^L \cdots \int dx_1 \cdots dx_N \exp\left[-\beta \sum_{1 \le i < j \le N} \phi(|x_i - x_j|)\right], \tag{2.1}$$

where here and henceforth we drop the trivial multiplicative factor λ^{-3N}.

For the Tonks gas

$$\phi(x) = \begin{cases} \infty & \text{if } |x| < a \\ 0 & \text{if } |x| > a, \end{cases} \tag{2.2}$$

and for the Takahashi gas [Takahashi (1942), see also Lieb and Mattis (B1966)]

$$\phi(x) = \begin{cases} \infty & \text{if } |x| < a \\ v(x-a) & \text{if } |x| > a \end{cases} \qquad \text{and } v(x-a) \equiv 0 \text{ if } |x| > 2a \tag{2.3}$$

The Tonks gas is simply a collection of hard rods of length a, and a moment's reflection shows that the Takahashi gas is a collection of hard rods with nearest-neighbor interactions only.

For general interaction potentials $\phi(x)$ the integral in Equation 2.1 is $N!$ times the integral over the region $R\colon 0 < x_1 < x_2 < \cdots < x_N < L$. (This is proved easily by induction using the result that the integrand is a symmetric function of the variables $x_1, x_2, \ldots, x_N$.) It is this fact and this fact alone that makes exact solutions in one dimension possible. Unfortunately, there is no analogue of this manipulation in higher dimensions.

Let us consider first the Tonks gas. From Equation 2.2 it follows that

$$\exp\left[-\beta \sum_{1 \le i < j \le N} \phi(|x_i - x_j|)\right] = \prod_{1 \le i < j \le N} S(|x_i - x_j|), \tag{2.4}$$

where

$$S(|x|) = \begin{cases} 1 & \text{if } |x| > a \\ 0 & \text{if } |x| < a. \end{cases} \tag{2.5}$$

The region of integration is therefore $|x_i - x_j| > a$ for all x_i and x_j. Moreover, in the region R it is obvious that the integrand equation (2.4) can be restricted to a product of terms with $j = i + 1$. The region of integration is then

$$R': 0 \leq x_1 < x_2 - a; a < x_2 < x_3 - a; \cdots (j-1)a < x_j < x_{j+1} - a; \\ \cdots (N-1)a < x_N \leq L;$$

hence from Equations 2.1 and 2.4,

$$Z(L, N, T) = \int_{(N-1)a}^{L} dx_N \int_{(N-2)a}^{x_N - a} dx_{N-1} \cdots \int_{a}^{x_3 - a} dx_2 \int_{0}^{x_2 - a} dx_1. \tag{2.6}$$

Changing variables to $y_j = x_j - (j-1)a$ then gives

$$\begin{aligned} Z(L, N, T) &= \int_0^l dy_N \int_0^{y_N} dy_{N-1} \cdots \int_0^{y_3} dy_2 \int_0^{y_2} dy_1 \\ &= \frac{l^N}{N!}, \end{aligned} \tag{2.7}$$

where

$$l = L - (N-1)a. \tag{2.8}$$

In the thermodynamic limit the free energy per particle ψ is given by

$$\begin{aligned} -\frac{\psi}{kT} &= \lim_{\substack{N, L \to \infty \\ v = L/N \text{ fixed}}} N^{-1} \log Z(L, N, T) \\ &= \lim_{\substack{N, L \to \infty \\ v = L/N \text{ fixed}}} N^{-1} \log\left(\frac{N^N}{N!}\right) + \log(v - a) \\ &= 1 + \log(v - a), \end{aligned} \tag{2.9}$$

which is a completely analytic function of v for $v > a$ (i.e., for densities $\rho < a^{-1}$, where it is to be noted that a^{-1} is the "close-packing density"). That is, for $\rho > a^{-1}$, at least two rods must overlap; hence from Equation 2.1, $Z(L, N, T) = 0$ when $\rho = N/L > a^{-1}$.

For the Takahashi gas we can repeat the above argument to Equation 2.6, obtaining

$$Z(L, N, T) = \int_{(N-1)a}^{L} dx_N \int_{(N-2)a}^{x_N - a} dx_{N-1} \cdots \int_0^{x_2 - a} dx_1 \exp\left[-\beta \sum_{j=1}^{N-1} \phi(x_{j+1} - x_j)\right], \quad (2.10)$$

since, from Equation 2.3, $\phi(|x_j - x_i|)$ vanishes in R' unless $j = i + 1$. In terms of the y_i variables (Equation 2.7) $x_{j+1} - x_j = y_{j+1} - y_j + a$; hence from Equation 2.3,

$$Z(L, N, T) = \int_0^l dy_N \int_0^{y_N} dy_{N-1} \cdots \int_0^{y_2} dy_1 \exp\left[-\beta \sum_{j=1}^{N-1} v(y_{j+1} - y_j)\right]. \quad (2.11)$$

We now define the convolution $f * g$ of two functions f and g by

$$f * g = \int_0^t f(t - x) g(x)\, dx. \quad (2.12)$$

It is a trivial exercise to verify from Equation 2.11 that

$$Z(L, N, T) = 1 * f * f * \cdots * f * 1, \quad (2.13)$$

where there are $N - 1$ f's occurring in the convolution and f is defined by

$$f(x) = \exp[-\beta v(x)]. \quad (2.14)$$

The Laplace convolution formula (see Problem 1) states that if

$$\mathscr{L}(f) = \int_0^\infty e^{-sx} f(x)\, dx \quad (2.15)$$

is the Laplace transform of $f(x)$, then

$$\mathscr{L}(f * g) = \mathscr{L}(f)\, \mathscr{L}(g). \quad (2.16)$$

Using this result we obtain from Equation 2.13, recalling Equation 2.11, that

$$\int_0^\infty e^{-ls} Z(L, N, T)\, dl = s^{-2}[K(s)]^{N-1}, \quad (2.17)$$

where

$$K(s) = \int_0^\infty \exp[-sx - \beta v(x)]\, dx \quad (2.18)$$

and we have used the facts that $\mathscr{L}(1) = s^{-1}$ and that Z, from Equation 2.11, can be considered as a function of $l = L - (N - 1)a$, N, and T. Further properties of the Takahashi gas which follow from equation 2.17 (e.g., that the gas does not undergo a phase transition) are left as exercises [see Problem 2 and Lieb and Mattis (B1966)].

As we have remarked, Van Hove (1950) proved that a one-dimensional gas of hard rods with a finite-range interaction does not have a phase transition [i.e., the limiting free energy $f(v, t)$ is an analytic function of v and T]. The Laplace convolution trick used for the Takahashi gas clearly does not work if the interaction involves more than nearest-neighbor rods. By a beautiful piece of analysis, however, Van Hove showed that for large N the problem can be reduced to finding the largest eigenvalue of a Fredholm integral equation in $\nu - 1$ variables, where ν is the number of interacting rods [i.e., $\phi(|x|) = 0$ if $|x| > (\nu + 1)a$]. Van Hove did not solve this problem but showed that the limiting free energy is a completely analytic function of v and T.

There has been considerable interest recently in finding under what conditions one can have a phase transition in one dimension. Ruelle and co-workers [see Ruelle (B1969)] succeeded in extending Van Hove's result by showing that there is no phase transition of the type shown in Figure 4.1 for a gas of hard rods with attractive interaction if

$$\int_0^R x|\phi(x - a)|\, dx < \infty \qquad \text{for all } R. \tag{2.19}$$

They did not prove complete analyticity, only that the pressure and its first derivative are continuous functions of v and T. This is sufficient to exclude a normal liquid–gas transition (Figure 4.1) but not a higher-order type of transition.

For a particular class of lattice systems, Dyson (1969a,b) has shown that if the integral in Equation 2.19 diverges sufficiently fast (essentially faster than log R), there is indeed a phase transition. He also extended Ruelle's result by showing that if the integral 2.19 does not diverge faster than log log R, there is no phase transition of the usual type. The interesting case of an inverse-square potential [i.e., $\phi(x) = x^{-2}$] is still, at the time of writing, undecided.

One final point to notice is that for the purely repulsive hard-core potential in any dimension (see Equations 2.2 and 2.4)

$$Z(V, N, T) = (N!)^{-1} \int_V \cdots \int d\mathbf{r}_1 \cdots d\mathbf{r}_N \prod_{1 \le i < j \le N} S(|\mathbf{r}_i - \mathbf{r}_j|), \tag{2.20}$$

where $S(|r|)$ is defined by Equation 2.5. It follows that the partition function does not depend on temperature, so if a phase transition exists (in dimension higher than one), it can only occur as a function of density. It is obvious that $Z(V, N, T)$ has a singularity at the close-packing density $\rho_0 = N_{max}/V$, where N_{max} is the maximum number of (hard) spheres of radius a that can be contained in the volume V. This is because the integrand in Equation 2.20 vanishes and hence

$$Z(V, N, T) = 0 \qquad \text{when } \rho > \rho_0. \tag{2.21}$$

Kirkwood conjectured some time ago that this system would have a phase transition at a particular density $\rho_c < \rho_0$, but this is still an unsolved problem. Note that if the Kirkwood transition exists, it would correspond to a type of gas–solid transition rather than a gas–liquid transition, since there is no temperature dependence. Dobrushin has recently shown that a two-dimensional lattice gas of hard squares does indeed have a phase transition, so it now appears extremely likely that a Kirkwood transition exists. Various numerical experiments [Alder and Wainwright (1962)] also support this view.

Additional existence theorems for phase transitions are known, but we postpone discussion of these until Section 4-4. In the following section we discuss the Yang–Lee characterization of a phase transition.

4-3 The Yang–Lee Characterization of a Phase Transition

An approach to the problem of phase transitions of a general nature was proposed by Yang and Lee in 1952 [Yang and Lee (1952b) and Lee and Yang (1952)]. Although it looked promising at the time, very little progress has been made since 1952. The basic idea is very simple, however, and perhaps with some clever insight something of value may emerge, so with this in mind we now present a brief discussion of the Yang–Lee theory.

The essential point of the argument is contained in Equation 2.21. Thus if we consider potentials having a finite hard core of radius a with no restrictions on the potential for $r > a$ except those necessary to ensure the existence of the thermodynamic limit (e.g. Equations 6.14 and 6.15 of Chapter 3) and denote by $M(V)$ the maximum number of spheres of radius a which can be contained in a (given) domain with volume V, then

$$Z(V, N, T) = 0 \qquad \text{for } N > M(V). \tag{3.1}$$

The grand-canonical partition function

$$\begin{aligned} Z_G(V, z, T) &= \sum_{N=0}^{\infty} z^N Z(V, N, T) \\ &= \sum_{N=0}^{M(V)} z^N Z(V, N, T) \end{aligned} \tag{3.2}$$

is then a polynomial of degree $M(V)$ in the fugacity z, and we can write [with the convention $Z(V, 0, T) = 1$]

$$Z_G(V, z, T) = \prod_{r=1}^{M(V)} \left(1 - \frac{z}{z_r}\right), \tag{3.3}$$

where z_r, $r = 1, 2, \ldots, M(V)$ are the (complex) zeros of $Z_G(V, z, T)$. Note that since all coefficients in Equation 3.2 are positive, there are no real positive roots, which is to say that there is no phase transition for finite V (and N).

The idea, then, is to see how a phase transition can develop in the limit of infinite V. Clearly $M(V) \to \infty$ when $V \to \infty$ and if, as is possible, the zeros coalesce into lines C in the complex plane, the grand-canonical potential $\chi(z, T)$ will have the form

$$\begin{aligned} \chi(z, T) &= \lim_{V \to \infty} V^{-1} \sum_{r=1}^{M(V)} \log\left(1 - \frac{z}{z_r}\right) \\ &= \int_C g(s) \log\left(1 - \frac{z}{s}\right) ds, \end{aligned} \tag{3.4}$$

where $Vg(s)\, ds$ is the number of zeros in a line element ds at z, i.e.,

$$\int_C g(s)\, ds = \lim_{V \to \infty} \frac{M(V)}{V}. \tag{3.5}$$

It is also conceivable that, in the limit, the zeros will coalesce into lines that "pinch" the positive real z axis, as shown in Figure 4.2. In such a case $\chi(z, T)$ will have two analytic pieces along the positive z axis, one to the left of AB and one to the right of AB. At z_c, then, $\chi(z, T)$ will be continuous in z but $\partial\chi(z, T)/\partial z$ will be discontinuous in z. Recall now from the definitions of p and v in the grand-canonical ensemble (Equations 5.2 and 5.3 of Chapter 3) that in the limit $V \to \infty$,

$$\beta p = \chi(z, T) \tag{3.6}$$

and

$$\rho = v^{-1} = z \frac{\partial}{\partial z} \chi(z, T). \tag{3.7}$$

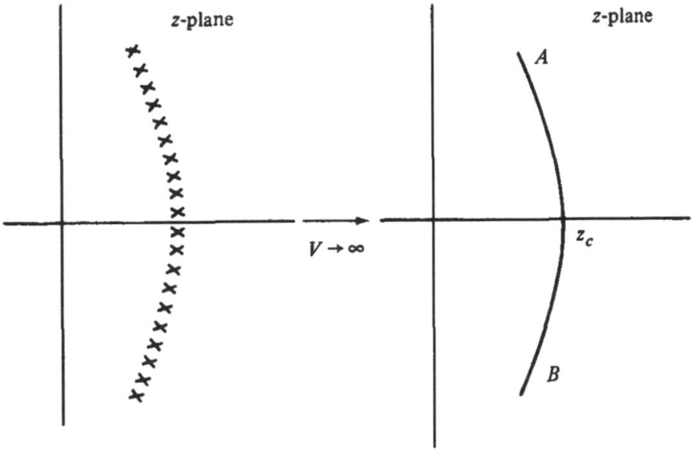

FIGURE 4.2. Distribution of zeros of the grand-canonical partition function in the complex z plane for a finite and an infinite system.

The equation of state is then obtained from these equations by eliminating the fugacity z. Thus, from the above remarks χ and $z\chi'$ as functions of z will be as shown in Figure 4.3 for a fixed temperature T. Eliminating z we get the typical isotherm shown in Figure 4.4.

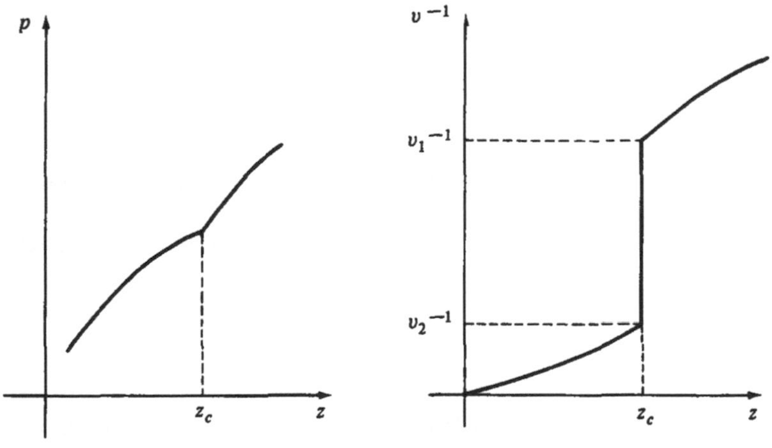

FIGURE 4.3. Pressure p and density v^{-1} as functions of (real) fugacity z.

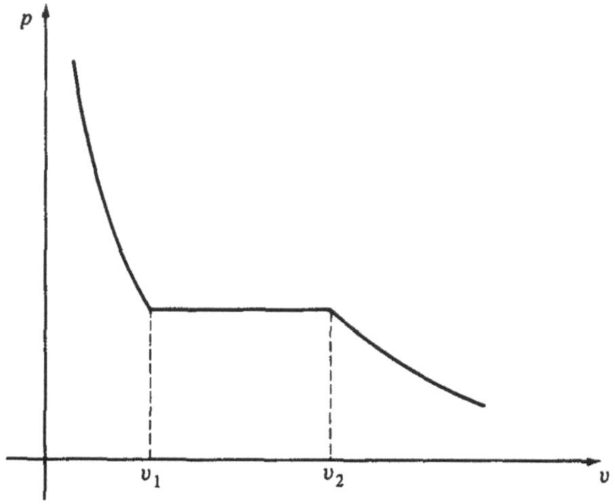

FIGURE 4.4 Isotherm obtained from Figure 4.3 by eliminating the fugacity z.

The Yang–Lee approach gives us a very nice characterization of a phase transition and demonstrates clearly how singular behavior might be approached as we take the limit of an infinite system. It still does not tell us, however, what we really want to know: Under what conditions on the potential $\phi(r)$ and dimensionality do we have a phase transition? At the present time, then, the Yang–Lee theory is only a characterization of a phase transition and is therefore a long way from being a complete theory.

In the next two sections we present the classical theories of phase transitions and discuss their range of validity. Some recent developments are discussed in Sections 4-6 and 4-7.

4-4 The Classical van der Waals Theory of Gas–Liquid Transitions

In his dissertation of 1873 van der Waals wrote down the following equation of state, which now bears his name:

$$\beta p = \frac{\rho}{1 - \rho v_0} - \frac{1}{2}\beta\alpha\rho^2. \tag{4.1}$$

v_0 is the "volume" of a particle ($\frac{4}{3}\pi r_0^3$ in three dimensions), so $\rho_0 = v_0^{-1}$ is the close-packing density, α is usually expressed in terms of the attractive part of the potential $\phi_{\text{att}}(r)$ by

$$\alpha = -\int \phi_{att}(r)\, d\mathbf{r}, \tag{4.2}$$

and the complete potential is usually written as

$$\phi(r) = \phi_{hc}(r) + \phi_{att}(r), \tag{4.3}$$

where $\phi_{hc}(r)$ is the hard core:

$$\phi_{hc}(r) = \begin{cases} \infty & r < r_0 \\ 0 & r > r_0 . \end{cases} \tag{4.4}$$

Equation 4.1 is usually written in terms of the specific volume v as

$$(v - v_0)\left(p + \frac{\alpha}{2v^2}\right) = kT. \tag{4.5}$$

Typical isotherms obtained from Equation 4.5 are shown in Figure 4.5.

Note that for constant p and T Equation 4.5 is a cubic equation for v, with roots v_1, v_2, and v_3, as shown in Figure 4.5. At the critical point, characterized by p_c, v_c, and T_c, the cubic equation for v_c must become

$$(v - v_c)^3 = v^3 - 3v^2 v_c + 3vv_c^2 - v_c^3 = 0. \tag{4.6}$$

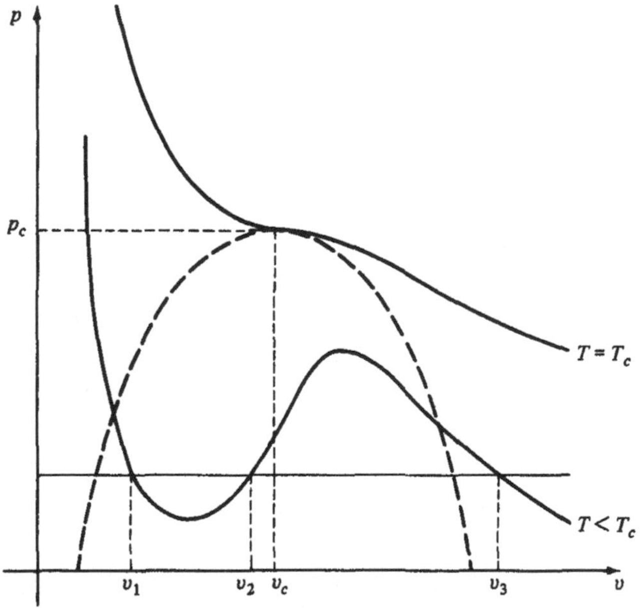

FIGURE 4.5. van der Waals isotherms obtained from Equation 4.5. For constant p and T Equation 4.5 is a cubic equation for v with typical roots v_1, v_2, and v_3. At the critical point characterized by p_c, v_c, T_c, $v_1 = v_2 = v_3 = v_c$.

But from Equation 4.5 we have that

$$(v - v_0)\left(p_c + \frac{\alpha}{2v^2}\right) = kT_c,$$

i.e.,

$$v^3 - \left(v_0 + \frac{kT_c}{p_c}\right)v^2 + \frac{\alpha}{2p_c}v - \frac{\alpha v_0}{2p_c} = 0, \tag{4.7}$$

which must be equivalent to Equation 4.6. Equating coefficients then gives the critical parameters

$$kT_c = \frac{4\alpha}{27v_0},$$

$$p_c = \frac{\alpha}{54v_0^2}, \tag{4.8}$$

$$v_c = 3v_0.$$

When $T < T_c$ the isotherms clearly violate the "stability condition" $\partial p/\partial v \leq 0$, which was shown in Section 3-7 to be a necessary consequence of statistical mechanics. Thermodynamically, this violation was realized almost immediately and Maxwell (1874) proposed the equal-area construction (i.e., area A = area B for the van der Waals isotherm, as shown in Figure 4.6) to remedy the difficulty.

From the definition of pressure p in terms of the free energy $f(v, T)$ (i.e., $kTp = \partial f/\partial v$), the "van der Waals wiggle" in the isotherm corresponds to a "kink" (indicated by a dashed line) in the free energy $f(v, T)$, as shown in Figure 4.7. The free energy is therefore not a convex function of v as required by statistical mechanics. It is not difficult to show, however (see Problem 5), that the equal-area construction on the isotherms is equivalent to the *double-tangent construction* shown in Figure 4.7. The modified free energy, shown as a solid line, is convex; in fact, it is the *convex envelope* of the van der Waals free energy (i.e., the minimal convex function, which is not less than the original function).

van der Waals, and many others since, thought that the wiggles or loops explained "metastable states" (i.e., the supercooled and supersaturated states shown in Figure 4.8). These states are observed experimentally, but, as we have noted, they cannot be derived from the partition function *after the*

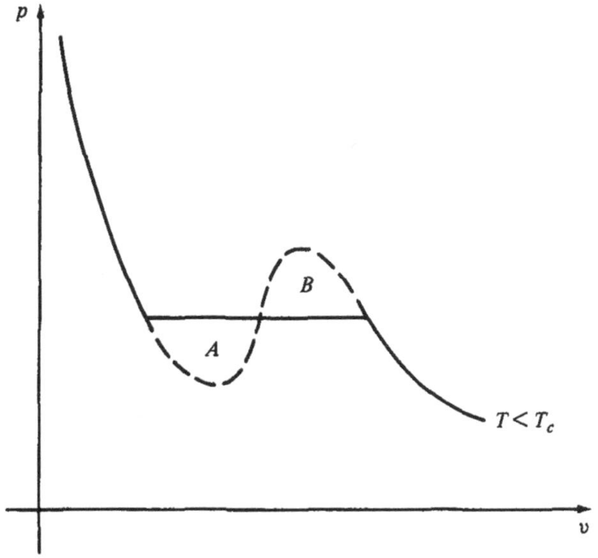

FIGURE 4.6. Maxwell equal-area construction. For the original van der Waals isotherm, area A is equal to area B.

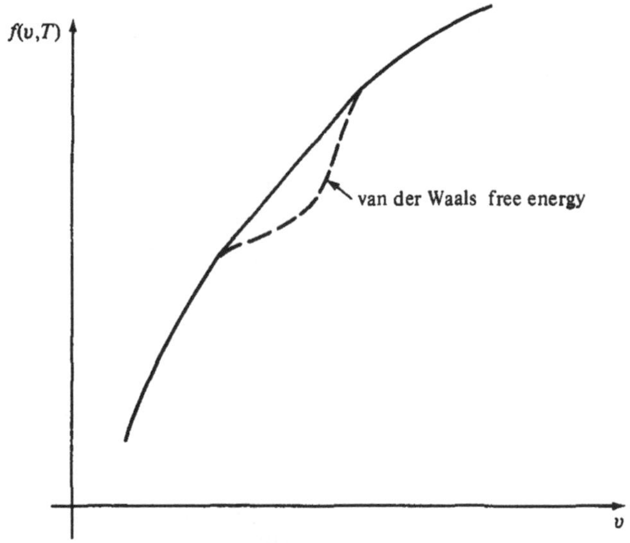

FIGURE 4.7. Double-tangent construction on the van der Waals free energy $f(v, T)$ as a function of v for fixed T. The solid curve is the convex envelope of the van der Waals free energy.

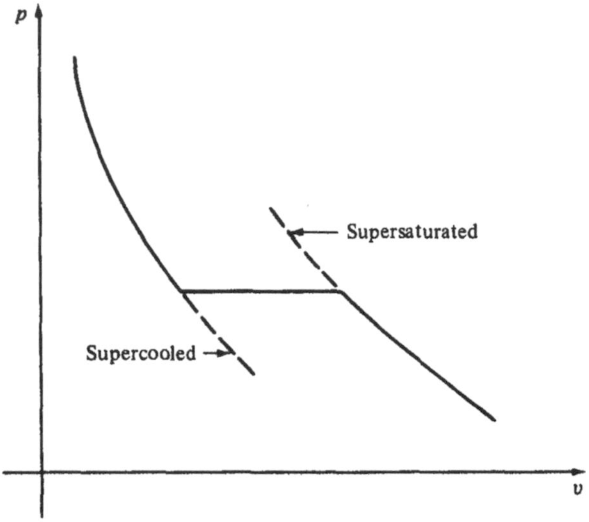

FIGURE 4.8. Supercooled and supersaturated metastable states.

thermodynamic limit has been taken. There is, as yet, no satisfactory theory of these states.

In spite of the "loop dilemma," the van der Waals equation with Maxwell construction enjoyed considerable success, particularly when it came to comparing the theory with experiment. Even today, with very refined experiments, the van der Waals theory gives a very good experimental fit away from the critical point. We will see in a moment, however, that the theory is grossly inadequate in the immediate neighborhood of the critical point. One would, nevertheless still like to understand the classical theory, particularly its apparent validity away from the critical point. This was first attempted (heuristically) by Ornstein in his dissertation of 1908. The essential idea is that the theory is valid in the limit of an infinitely long range attractive potential (with superimposed hard core). The argument goes roughly as follows: With interaction 4.3 the partition function is given by (see Equation 2.4)

$$Z(V, N, T) = (N!)^{-1} \int_V \cdots \int d\mathbf{r}_1 \cdots d\mathbf{r}_N \prod_{1 \le i < j \le N} S(|\mathbf{r}_i - \mathbf{r}_j|) \times \exp\left[-\beta \sum_{1 \le i < j \le N} \phi_{\text{att}}(|\mathbf{r}_i - \mathbf{r}_j|)\right], \quad (4.9)$$

where $S(|\mathbf{r}|)$ is the exponential of the hard-core part of the potential, i.e.,

$$S(r) = \begin{cases} 0 & r < r_0 \\ 1 & r > r_0 . \end{cases} \tag{4.10}$$

Now, Ornstein argues, if the range of $\phi_{\text{att}}(r)$ is very large, the potential energy for almost all configurations of the N particles will be equal to the average value, i.e.,

$$\sum_{1 \le i < j \le N} \phi_{\text{att}}(|\mathbf{r}_i - \mathbf{r}_j|) \approx -\frac{\alpha N^2}{2V} = -\frac{\alpha N \rho}{2} \tag{4.11}$$

for almost all configurations $\mathbf{r}_1, \ldots, \mathbf{r}_N$ in Equation 4.9, where

$$\alpha = -\int \phi_{\text{att}}(r)\, d\mathbf{r} \qquad \text{and} \qquad \rho = \frac{N}{V}. \tag{4.12}$$

If this is so,

$$Z(V, N, T) \approx \exp\left(\frac{\beta \alpha N \rho}{2}\right)\left[(N!)^{-1} \int_V \cdots \int d\mathbf{r}_1 \cdots d\mathbf{r}_N \prod_{1 \le i < j \le N} S(|\mathbf{r}_i - \mathbf{r}_j|)\right], \tag{4.13}$$

where the term in brackets is the partition function $Z^{\text{hc}}(V, N)$ for a gas of hard spheres of radius r_0. As we have noted, the evaluation of $Z^{\text{hc}}(V, N)$ is an unsolved problem for dimensionality higher than one. In one dimension (Equation 2.7)

$$Z^{\text{hc}}(V, N) = \frac{[V - (N-1)v_0]^N}{N!}, \tag{4.14}$$

where V is the one-dimensional volume (i.e., length L of the domain) and v_0 is the one-dimensional volume of each particle (i.e., r_0). If one believes that Equation (4.14) is a reasonable approximation to the truth in higher dimensions (taking $v_0 = \frac{4}{3}\pi r_0^3$ in three dimensions, etc.), then one has, in the thermodynamic limit,

$$-\frac{\psi}{kT} = \lim_{\substack{N, V \to \infty \\ \rho = N/V \text{ fixed}}} N^{-1} \log Z(V, N, T) = 1 + \frac{\beta \alpha \rho}{2} + \log(v - v_0), \tag{4.15}$$

which gives

$$p = -\frac{\partial \psi}{\partial v} = kT[(v - v_0)^{-1} - \tfrac{1}{2}\beta \alpha \rho^2], \tag{4.16}$$

or, if one does not make use of Equation 4.14,

$$\frac{p}{kT} = \frac{p^{\text{hc}}}{kT} - \frac{1}{2}\beta\alpha\rho^2, \tag{4.17}$$

where p^{hc} is the hard-core pressure [i.e., $kT(v - v_0)^{-1}$ in one dimension but strictly unknown in higher dimensions].

Since we have ended up with the van der Waals equation of state (4.16) including loops, which we know cannot result from the partition function, we have obviously committed an error somewhere. The difficulty can be traced back to Equation 4.11, where we have let the effective range of interaction depend on the size of the system. This violates one of the basic principles of statistical mechanics: The interaction potential cannot depend explicitly on the size of the system. One should really let the range of interaction go to infinity *after* the thermodynamic limit has been taken. This is the key to a recent result of Lebowitz and Penrose (1966), which is as follows.

Let, for dimensionality d,

$$\phi_{\text{att}}(r) = \gamma^d v(\gamma r), \tag{4.18}$$

where $v(r)$ is sufficiently regular to ensure the existence of the thermodynamic limit (see Equations 6.14 and 6.15 of Chapter 3). The γ scales the potential while preserving α; i.e.,

$$\begin{aligned}\alpha &= -\int \phi_{\text{att}}(r)\, d\mathbf{r} \\ &= -\gamma^d \int_0^\infty r^{d-1}\, dr v(\gamma r) \int d\sigma_d \\ &= -C_d \int_0^\infty x^{d-1} v(x)\, dx\end{aligned} \tag{4.19}$$

is independent of γ, where $C_d = \int d\sigma_d$ is the d-dimensional solid angle. For small γ the potential $\phi_{\text{att}}(r)$ is very weak and very flat, so in a sense γ is a measure of the inverse range of interaction. By a construction that is reminiscent of the cube construction in the proof of the existence of the thermodynamic limit, Lebowitz and Penrose obtained upper and lower bounds for $Z(V, N, T)$, and hence for the free energy $\Psi(V, N, T)$, which in the thermodynamic limit followed by the limit $\gamma \to 0$ (i.e., infinitely weak long-range attraction) coalesce to give for the free energy per particle ψ:

$$\frac{\psi(\rho, T)}{kT} = \text{convex envelope}\left[\frac{\psi^{\text{hc}}(\rho, T)}{kT} - \tfrac{1}{2}\beta\alpha\rho\right], \tag{4.20}$$

which is the van der Waals free energy together with the Maxwell or double-tangent construction, as explained above. Equation 4.20 gives the van der Waals equation of state,

$$\frac{p}{kT} = \frac{p^{\text{hc}}}{kT} - \frac{1}{2}\beta\alpha\rho^2 \qquad \textit{with the Maxwell construction.} \tag{4.21}$$

The hard-core free energy $\psi^{\text{hc}}(\rho, T)$, as we have noted, is unknown in dimensions higher than one, but as far as isotherms are concerned this regrettable fact is immaterial, since $\psi^{\text{hc}}(\rho, T)/kT$ is independent of temperature.

In Appendix C we have given the lattice version of the Lebowitz–Penrose proof. This is appropriate for lattice spin or gas systems discussed in the following section (so it is suggested that the interested reader complete Section 4-5 before turning to Appendix C).

Historically, Equation 4.20 was first derived for a one-dimensional gas of hard rods with $v(x)$ (Equation 4.18) equal to $\exp(-|x|)$, by Kac et al. (1963), Uhlenbeck et al. (1963), and Hemmer et al. (1964). Guided by this work, Van Kampen (1964) extended Ornstein's heuristic argument to derive Equation 4.20, but the final proof was given first by Lebowitz and Penrose.

Although we now understand the validity of the van der Waals theory, we note that the final result is dimension independent [dimensionality sits in $\psi^{\text{hc}}(\rho, T)$, which has a trivial temperature dependence], and, as we will see shortly, the most interesting and perhaps challenging features of phase transitions are intimately connected in some mysterious and unknown way with dimensionality.

4-5 The Classical Curie–Weiss Theory of Magnetic Transitions

In many ways the theoretical discussion of magnetic transitions is simpler than the corresponding transition for gaseous systems. Thermodynamically the descriptions are almost identical, but there are simplifying features for magnets.

Corresponding to the gas–liquid isotherms (Figure 4.1) we have the magnetic isotherms shown in Figure 4.9. Experimentally, the situation is as follows. If a magnetic field H is applied to a magnet (in a particular direction) at sufficiently low temperatures T ($T < T_c$) the magnet develops a magnetization $M(H, T)$. If the field is then turned off, a *residual* or *spontaneous magnetization* $M_0(T) = \lim_{H\to 0+} M(H, T)$ remains as long as $T < T_c$. When

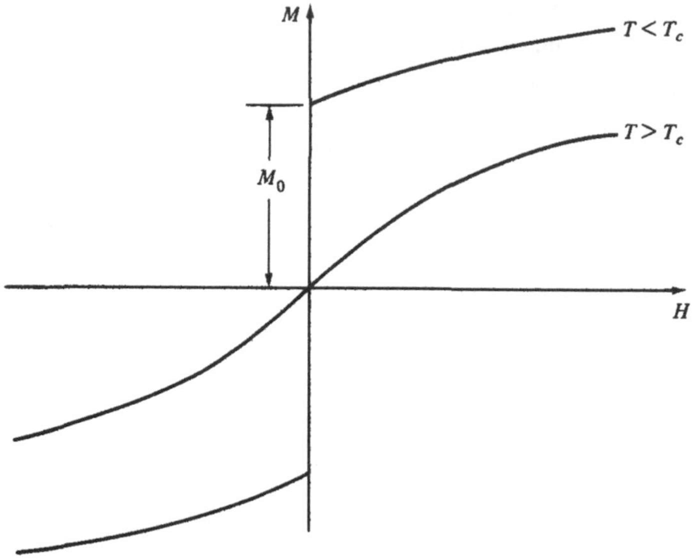

FIGURE 4.9. Magnetic isotherms: magnetization M as a function of magnetic field H at constant temperature T. M_0 is the spontaneous or residual magnetization.

$T \geq T_c$, there is no spontaneous magnetization and M_0 vanishes abruptly at $T = T_c$, as shown in Figure 4.10.

Comparison with Figure 4.1 shows that the magnetic isotherms are like gas–liquid isotherms turned on their side and flattened. Magnetization M then corresponds to volume V and the negative of the magnetic field, $-H$, to the pressure P. Thermodynamic relations for magnetic systems can then be obtained directly from Chapter 2 by replacing V by M and P by $-H$. The simplifying feature for magnetic transitions is, of course, the fact that the flat parts of the isotherms always lie on the M axis.

A model of a magnetic system is typically composed of a set of N spins $\mu_i, i = 1, 2, \ldots, N$, which may be vectors, scalars, or, in the quantum-mechanical case, spin operators. We shall consider the simplest system composed of scalar spins $\mu_i = +1$ or -1 corresponding to spin up or spin down, respectively. The interaction energy between two spins μ_i and μ_j located at *fixed points* $\mathbf{r}_i$ and $\mathbf{r}_j$ in space, respectively (e.g., on the vertices of a regular lattice) is taken to be $+\phi(|\mathbf{r}_i - \mathbf{r}_j|)$ if the spins are parallel ($\mu_i \mu_j = +1$) and $-\phi(|\mathbf{r}_i - \mathbf{r}_j|)$ if the spins are antiparallel ($\mu_i \mu_j = -1$). In other words, the

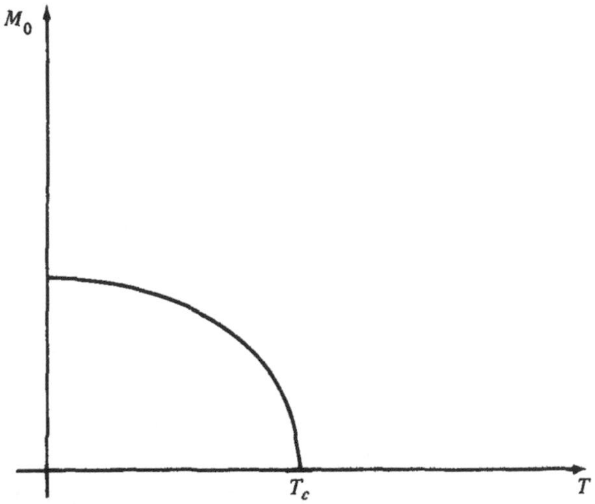

FIGURE 4.10. Spontaneous magnetization M_0 as a function of temperature T. M_0 vanishes abruptly at $T = T_c$ and is zero for $T \geq T_c$.

total interaction energy of the system in a given "configuration" $\{\mu\} = (\mu_1, \mu_2, \ldots, \mu_N)$ of spins is given by

$$E\{\mu\} = \sum_{1 \leq i < j \leq N} \phi(|\mathbf{r}_i - \mathbf{r}_j|)\mu_i \mu_j - H \sum_{i=1}^{N} \mu_i, \tag{5.1}$$

where the first term comes from interactions between spins and the second term from interactions of each spin with the magnetic field H. Notice that when $\phi(r) \leq 0$ we get the lowest possible energy (the ground state) for a parallel configuration of spins, e.g., when $H > 0$,

$$\min_{\{\mu\}} E\{\mu\} = \sum_{1 \leq i < j \leq N} \phi(|\mathbf{r}_i - \mathbf{r}_j|) - NH \qquad \text{when all } \mu_i = +1. \tag{5.2}$$

This corresponds to a *ferromagnetic system*. Similarly, when $\phi(r) \geq 0$, we get the lowest energy for an antialigned configuration of spins (which is not always easy to find!) corresponding to an *antiferromagnetic* system. We shall have more to say about this in Chapter 5.

Now since the positions $\mathbf{r}_i$ are asumed to be fixed, the partition function by analogy with Equation 4.20 of Chapter 3 is obtained by summing over all possible spin configurations (rather than integrating over positions of particles for a gas), i.e.,

$$Z(N, T) = \sum_{\{\mu\}} \exp(-\beta E\{\mu\}), \tag{5.3}$$

where the sum over $\{\mu\}$ denotes a sum over all configurations (2^N in total) $\mu_1 = \pm 1, \mu_2 = \pm 1, \ldots, \mu_N = \pm 1$.

In the particular case where the $\mathbf{r}_i$'s are lattice points, i.e., the spins μ_i are located on the vertices of a lattice (for example, a simple cubic lattice) and the interaction $\phi(|\mathbf{r}_i - \mathbf{r}_j|)$ is between nearest-neighbor lattice sites only, the system 5.1 is referred to as the *Ising model.* We shall have much more to say about this particular case in Chapter 5. Notice also that the system 5.1 can be considered as a model of a gas when the gas particles are only allowed to occupy discrete points $\mathbf{r}_i$ in space. Spin up would then correspond to, say, a particle at position $\mathbf{r}_i$ and spin down to no particle at $\mathbf{r}_i$. We shall discuss this interpretation also in Chapter 5.

Our immediate purpose here is to discuss the classical Curie–Weiss theory of ferromagnetism for the simple case of Equation 5.1. It is clear, first of all, that a general evaluation of the partition function 5.3 is out of the question. We are then left with either special cases, such as the Ising model, or an approximate evaluation of 5.3. The particular approximation that we now discuss is based on Ornstein's observation for the gas. We assume that the range of interaction is very long, so for most configurations each spin effectively interacts with a *mean field* produced by all the other spins. This means that for most configurations each spin has an effective constant interaction with all other spins; i.e., in the exponent of Equation 5.3 we make the approximation

$$\sum_{1 \le i < j \le N} \phi(|\mathbf{r}_i - \mathbf{r}_j|)\,\mu_i \mu_j \approx -\frac{\alpha}{N} \sum_{1 \le i < j \le N} \mu_i \mu_j, \tag{5.4}$$

where α, as before, is given by

$$\alpha = -\int \phi(r)\,d\mathbf{r} \ge 0. \tag{5.5}$$

The problem then is to evaluate the partition function (5.3) with interaction energy $E\{\mu\}$ given by

$$E\{\mu\} = -\frac{\alpha}{N} \sum_{1 \le i < j \le N} \mu_i \mu_j - H \sum_{i=1}^{N} \mu_i . \tag{5.6}$$

We can consider this modified problem as either an approximation to the original problem or as a special case of the original problem. The interaction 5.6, in fact, defines what is commonly referred to as the Curie–Weiss model. It should be noted that the interaction 5.6 depends explicitly on the number of spins, so it is really not a bona fide statistical-mechanical model. Notice also that the model is independent of dimension, so in all respects it is identical to the Ornstein approximation for the gas. Nevertheless, we can solve the Curie–Weiss model exactly and, as shown in Appendix C, the final result can be obtained rigorously by taking a potential $\phi(r)$ of the "Kac type," Equation 4.18, and allowing the range of interaction to approach infinity *after* the thermodynamic limit.

To evaluate the partition function 5.3 for the interaction 5.6 we first symmetrize the interaction by writing

$$E\{\mu\} = -\frac{\alpha}{2N}\left(\sum_{i=1}^{N}\mu_i\right)^2 + \frac{\alpha}{2} - H\sum_{i=1}^{N}\mu_i. \tag{5.7}$$

This follows from Equation 5.6 and the fact that $\mu_i^2 = 1$. The partition function 5.3 can then be written as

$$Z(N, T) = \exp\left(-\frac{\nu}{2}\right)\sum_{\{\mu\}}\exp\left[\frac{\nu}{2N}\left(\sum_{i=1}^{N}\mu_i\right)^2 + B\sum_{i=1}^{N}\mu_i\right], \tag{5.8}$$

where

$$\nu = \beta\alpha \qquad \text{and} \qquad B = \beta H. \tag{5.9}$$

We now use the elementary identity

$$\exp\left(\frac{a}{2}\right) = (2\pi)^{-1/2}\int_{-\infty}^{\infty}\exp(-\tfrac{1}{2}x^2 + \sqrt{a}x), \tag{5.10}$$

which is easily proved by completing the square in the exponent on the right-hand side and using the well-known result

$$\int_{-\infty}^{\infty}\exp\left(-\frac{y^2}{2}\right)dy = (2\pi)^{1/2}. \tag{5.11}$$

Taking $a^{1/2} = \sqrt{\nu/N}(\sum_{i=1}^{N}\mu_i)$ in Equation 5.10, the partition function 5.8 can be written as

$$Z(N, T) = \exp\left(-\frac{\nu}{2}\right)\sum_{\{\mu\}}(2\pi)^{-1/2}\int_{-\infty}^{\infty}\exp\left[-\frac{1}{2}x^2 + \left(x\sqrt{\frac{\nu}{N}} + B\right)\sum_{i=1}^{N}\mu_i\right]dx. \tag{5.12}$$

The spins are now completely uncoupled and since

$$\begin{aligned}\sum_{\mu=\pm 1} \exp(K\mu) &= \exp(K) + \exp(-K) \\ &= 2\cosh K,\end{aligned} \tag{5.13}$$

the sum over configurations $\{\mu\}$ in Equation 5.12 can be carried out immediately, to give

$$Z(N, T) = \exp\left(-\frac{\nu}{2}\right)(2\pi)^{-1/2} \int_{-\infty}^{\infty} \exp\left(-\frac{x^2}{2}\right)\left[2\cosh\left(x\sqrt{\frac{\nu}{N}} + B\right)\right]^N dx. \tag{5.14}$$

Changing variables to $\eta = x(\nu N)^{-1/2}$ we obtain

$$Z(N, T) = \exp\left(-\frac{\nu}{2}\right)2^N\left(\frac{\nu N}{2\pi}\right)^{1/2} \int_{-\infty}^{\infty} \left[\exp\left(-\frac{\nu\eta^2}{2}\right)\cosh(\nu\eta + B)\right]^N d\eta. \tag{5.15}$$

The integral in Equation 5.15 is of the form

$$I(N) = \int_{-\infty}^{\infty} \exp[Nf(\eta)]\, d\eta, \tag{5.16}$$

where

$$f(\eta) = -\frac{\nu\eta^2}{2} + \log\cosh(\nu\eta + B). \tag{5.17}$$

In the limit $N \to \infty$ (the thermodynamic limit) $I(N)$ can be evaluated asymptotically by Laplace's method, i.e.,

$$I(N) \sim C_N \exp\left[N \max_{-\infty<\eta<\infty} f(\eta)\right] \qquad \text{as } N \to \infty, \tag{5.18}$$

where C_N is such that $\lim_{N\to\infty} N^{-1}\log C_N = 0$. The free energy per spin ψ in the thermodynamic limit is therefore given by

$$\begin{aligned}-\frac{\psi}{kT} &= \lim_{N\to\infty} N^{-1}\log Z(N, T) \\ &= \log 2 + \max_{-\infty<\eta<\infty} f(\eta).\end{aligned} \tag{5.19}$$

The equation determining the maximum in 5.19 is, from Equation 5.17,

$$\frac{df}{d\eta} = -\nu\eta + \nu\tanh(\nu\eta + B) = 0, \tag{5.20}$$

i.e.,

$$\eta = \tanh(\nu\eta + B), \tag{5.21}$$

which is precisely the equation determining the mean field η in the Curie–Weiss theory.

The magnetization $M(H, T)$ is defined in general by

$$\begin{aligned} M(H, T) &= [Z(N, T)]^{-1} \sum_{\{\mu\}} \left(\sum_{i=1}^{N} \mu_i \right) \exp(-\beta E\{\mu\}) \\ &= \frac{\partial}{\partial B} [\log Z(N, T)]; \end{aligned} \tag{5.22}$$

hence the magnetization per spin $m(H, T)$ in the thermodynamic limit is given by

$$\begin{aligned} m(H, T) &= \lim_{N\to\infty} N^{-1} M(H, T) \\ &= -\frac{\partial}{\partial B} \frac{\psi}{kT}. \end{aligned} \tag{5.23}$$

From Equations 5.17, 5.19, and 5.20 it follows straightforwardly that

$$m(H, T) = \eta, \tag{5.24}$$

where η is the solution of Equation 5.21, which *maximizes* Equation 5.17. In the limit $H \to 0+$ we obtain the spontaneous magnetization

$$m_0 = \lim_{H\to 0+} m(H, T) = \eta_0, \tag{5.25}$$

where η_0 is the nonnegative solution of

$$\eta_0 = \tanh(\nu\eta_0). \tag{5.26}$$

The graphical solution of this equation is shown in Figure 4.11.

When the slope of $\tanh(\nu\eta)$ at the origin (i.e., ν) exceeds unity, we have three solutions of Equation 5.26, $\pm\eta_0$ and 0. An elementary calculation shows that the nonzero solutions $\pm\eta_0$ maximize $f(\eta)$, so m_0 is nonzero when $\nu = \alpha/kT > 1$. On the other hand, when $\nu \leq 1$, $\eta_0 = 0$ is the only solution of Equation 5.26; hence the spontaneous magnetization is zero when $\nu = \alpha/kT \leq 1$.

$$T_c = \frac{\alpha}{k} \tag{5.27}$$

therefore represents the critical or "Curie point," and

$$m_0 = \begin{cases} \eta_0 & \text{when } T < T_c \\ 0 & \text{when } T \geq T_c, \end{cases} \tag{5.28}$$

as shown in Figure 4.12.

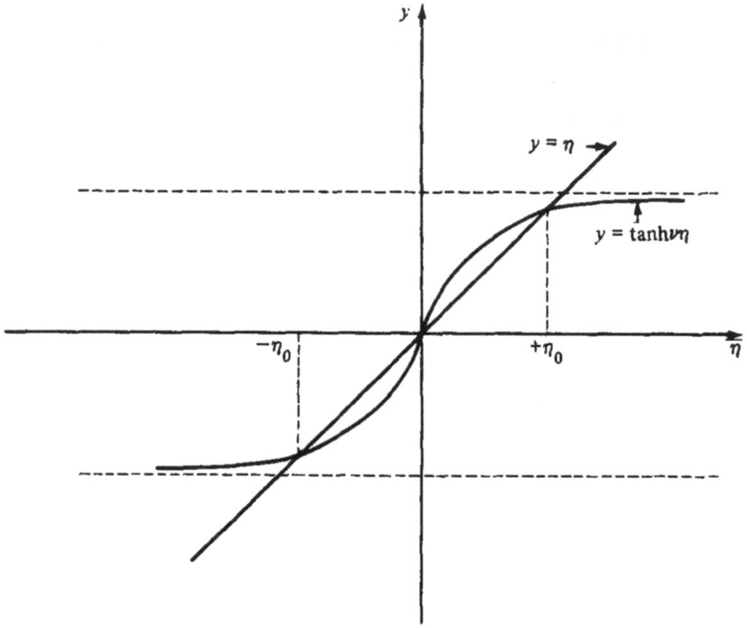

FIGURE 4.11. Graphical solution of the Curie–Weiss mean-field equation (Equation 5.26).

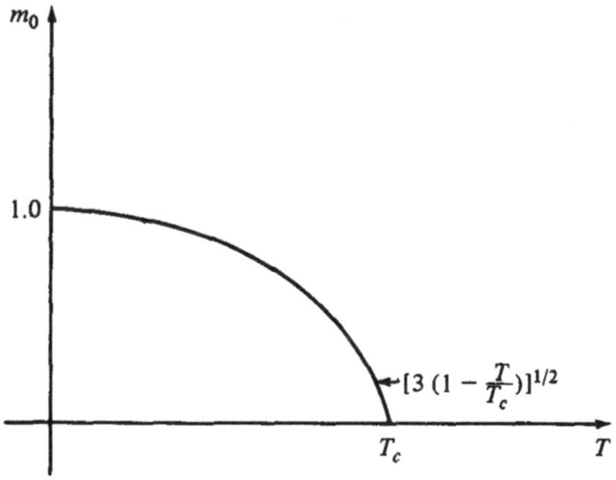

FIGURE 4.12. Spontaneous magnetization per spin in the Curie–Weiss theory, Equation 5.28.

In the neighborhood of $T = T_c-$, η_0 is small, so we can expand $\tanh(\nu\eta_0)$ in Equation 5.26 to get

$$\eta_0 \sim \nu\eta_0 - \frac{(\nu\eta_0)^3}{3} + \cdots \qquad \text{as } T \to T_c - \qquad (\text{i.e., } \nu \to 1+),$$

which gives

$$\eta_0 \sim \left[3\left(1 - \frac{T}{T_c}\right)\right]^{1/2} \qquad \text{as } T \to T_c -. \tag{5.29}$$

The zero-field free energy from Equations 5.19 and 5.21 is given by

$$-\frac{\psi_0}{kT} = \begin{cases} \log 2 & \text{when } T \geq T_c \\ \log 2 - \dfrac{\nu\eta_0^2}{2} + \log\cosh(\nu\eta_0) & \text{when } T < T_c, \end{cases} \tag{5.30}$$

where η_0 is the nonzero solution of Equation 5.26 (note that ψ_0 is a symmetric function of η_0). The zero-field specific heat is then given by

$$C = -T\frac{\partial^2\psi_0}{\partial T^2} = \begin{cases} 0 & \text{when } T > T_c \\ -\dfrac{\alpha}{2}\dfrac{d\eta_0^2}{dT} & \text{when } T < T_c, \end{cases} \tag{5.31}$$

which has the form shown in Figure 4.13, i.e., a jump discontinuity at $T = T_c$.

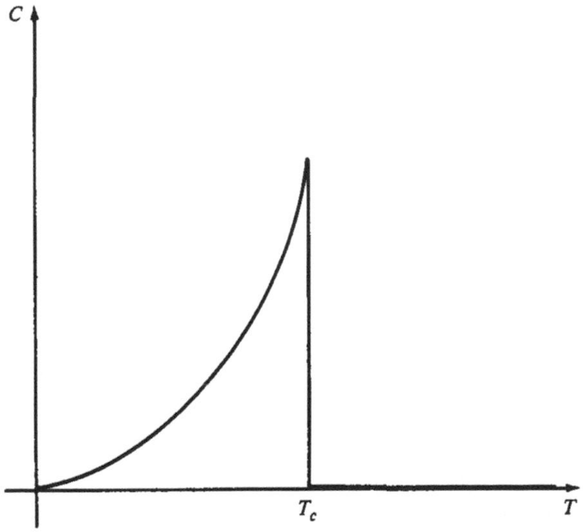

FIGURE 4.13. Zero-field specific heat C as a function of temperature T in the Curie–Weiss theory, Equation 5.31.

Another quantity of interest is the isothermal susceptibility χ (we suppress the temperature suffix), defined by

$$\chi = \frac{\partial M}{\partial H}. \tag{5.32}$$

The magnetization M is equal to η (Equation 5.21) and an elementary calculation shows that

$$\chi = \beta(1-\eta^2)[1-\nu(1-\eta^2)]^{-1}. \tag{5.33}$$

The zero-field susceptibility χ_0 is therefore given by

$$\chi_0 = \beta(1-\eta_0^2)[1-\nu(1-\eta_0^2)]^{-1}, \tag{5.34}$$

where η_0 is the nonzero solution of Equation 5.26. From Equation 5.29 and the fact that $\eta_0 = 0$ when $T \geq T_c$, it follows that in the neighborhood of $T = T_c$ (i.e., either $T \to T_c+$ or $T \to T_c-$)

$$\chi_0 \sim \left|1 - \frac{T}{T_c}\right|^{-1} \qquad \text{as } T \to T_c. \tag{5.35}$$

When H is nonzero the magnetic isotherms, from Equation 5.21, have the form shown in Figure 4.9. Of particular interest is the behavior of the $T = T_c$ isotherm as $H \to 0+$. In this neighborhood η (i.e., the magnetization) is small, so from Equations 5.21 and 5.27 we obtain

$$\eta \sim (\eta + \beta_c H) - \frac{(\eta + \beta_c H)^3}{3} + \cdots \qquad \text{at } T = T_c \text{ and } H \to 0+,$$

from which it follows that

$$\eta \sim H^{1/3} \qquad \text{at } T = T_c \text{ as } H \to 0+. \tag{5.36}$$

The behavior of thermodynamic quantities—such as magnetization, susceptibility, and specific heat at zero field—in the neighborhood of the critical point is of some interest, since recent experimental techniques enable us to examine the behavior of such quantities very close to the critical point. Unfortunately, the predictions of the classical theory, Equations 5.29, 5.31, 5.35, and 5.36 do not agree with experiment. We shall have more to say about this in the following section.

As we have previously remarked, Equation 5.30 for the zero-field free energy (or more generally Equation 5.19) is valid in the following sense.

Consider a potential $\phi(r)$ in Equation 5.1 of the "Kac-type" equation 4.18, i.e.,

$$\phi(r) = \gamma^d v(\gamma r) \qquad \text{in } d \text{ dimensions.} \tag{5.37}$$

If we take the limit $\gamma \to 0$, corresponding to an infinitely weak long-range interaction, after the thermodynamic limit, then the free energy is given by Equation 5.30. A slightly stronger version of this theorem, which is a straightforward extension of the Lebowitz–Penrose theorem stated in the previous section, is proved in Appendix C.

4-6 Critical-point Exponents

We have seen in the previous section that the classical Curie–Weiss theory of magnetism predicts the following behavior in the neighborhood of the critical point:

$C_{H=0}$ has a jump discontinuity at $T = T_c$

$$m_0 \sim \left(1 - \frac{T}{T_c}\right)^{1/2} \qquad \text{as } T \to T_c-$$

$$\chi_0 \sim \left|1 - \frac{T}{T_c}\right|^{-1} \qquad \text{as } T \to T_c \tag{6.1}$$

and

$$m \sim H^{1/3} \qquad \text{at } T = T_c \text{ as } H \to 0+,$$

where $C_{H=0}$ is the zero-field specific heat, m_0 the spontaneous magnetization, χ_0 the zero-field susceptibility, and m the magnetization. The power $\frac{1}{2}$ for m_0 and -1 for χ_0 are particular examples of critical exponents, which are defined in general as follows: *The critical exponent λ corresponding to the critical point $x = 0+$ of a function $f(x)$ is defined by*

$$\lambda = \lim_{x \to 0+} \frac{\log f(x)}{\log x}. \tag{6.2}$$

We write

$$f(x) \sim x^\lambda \qquad \text{as } x \to 0+ \tag{6.3}$$

if the limit (6.2) *exists.*

It is to be noted that a function with a jump discontinuity at $x = 0$ has exponent $\lambda = 0$, as does the function $\log x$. We must be careful, therefore, to distinguish, among other things, an exponent 0 for a discontinuity and an exponent 0 for a logarithmic divergence. We shall usually write, except when it is obvious, 0_{disc} and $0_{\log}$ for these two cases, respectively.

The standard definitions of critical exponents are given for a number of thermodynamic quantities in Table 4.1. The list is by no means exhaustive

TABLE 4.1. Definition of some critical exponents for fluid and magnetic systems (see the text).

Exponent	Fluid	Magnet
α'	$C_V \sim \left(1-\dfrac{T}{T_C}\right)^{-\alpha'}$ as $T \to T_C-$	$C_{H=0} \sim \left(1-\dfrac{T}{T_C}\right)^{-\alpha'}$ as $T \to T_C-$
α	$C_V \sim \left(1-\dfrac{T_C}{T}\right)^{-\alpha}$ as $T \to T_C+$	$C_{H=0} \sim \left(1-\dfrac{T_C}{T}\right)^{-\alpha}$ as $T \to T_C+$
β	$\rho_L - \rho_G \sim \left(1-\dfrac{T}{T_C}\right)^{\beta}$ as $T \to T_C-$	$m_0 \sim \left(1-\dfrac{T}{T_C}\right)^{\beta}$ as $T \to T_C-$
γ'	$K_T \sim \left(1-\dfrac{T}{T_C}\right)^{-\gamma'}$ as $T \to T_C-$	$\chi_0 \sim \left(1-\dfrac{T}{T_C}\right)^{-\gamma'}$ as $T \to T_C-$
γ	$K_T \sim \left(1-\dfrac{T_C}{T}\right)^{-\gamma}$ as $T \to T_C+$	$\chi_0 \sim \left(1-\dfrac{T_C}{T}\right)^{-\gamma}$ as $T \to T_C+$
δ	$p - p_C \sim \text{sgn}(\rho_L - \rho_G)\lvert\rho_L - \rho_G\rvert^{\delta}$ at $T = T_C$ as $\lvert\rho_L - \rho_G\rvert \to 0$	$H \sim \text{sgn}(m)\lvert m\rvert^{\delta}$ at $T = T_C$ as $H \to 0$

and the interested reader is referred to Fisher (1967) and Stanley (B1971) for a more complete list. The thermodynamic quantities in Table 4.1 are for the gas:

C_V = specific heat at constant volume

ρ_L, ρ_G = liquid and gas densities, respectively

K_T = isothermal compressibility

and the corresponding thermodynamic quantities for the magnet:

$C_{H=0}$ = zero-field specific heat

m_0 = spontaneous magnetization

χ_0 = zero-field isothermal susceptibility.

We have shown that for the Curie–Weiss theory

$$\alpha' = \alpha = 0_{\text{disc}}, \quad \beta = \tfrac{1}{2}, \quad \gamma' = \gamma = 1, \quad \text{and } \delta = 3. \tag{6.4}$$

It turns out that these exponents are the same for the van der Waals theory (see Problem 6). There is also close experimental agreement, as shown in Table 4.2, between magnetic and fluid systems.

TABLE 4.2. Comparison of experimental critical exponents with predictions of the classical theories.

Exponent	Classical Theories	Experimental	
		Fluid Systems	Magnetic Systems
α'	0_{disc}	$0_{\log} \lesssim \alpha' \lesssim 0.2$	$0_{\log} \lesssim \alpha' \lesssim 0.2$
α	0_{disc}	$0_{\log} \lesssim \alpha \lesssim 0.2$	$0_{\log} \lesssim \alpha \lesssim 0.2$
β	$\frac{1}{2}$	0.33	0.33
γ'	1	1.1	?
γ	1	1.3	1.3
δ	3	4.2	4.2

The magic numbers we have listed under "experimental" do not represent only one set of data. Rather, they represent a slightly biased "average" of values obtained for a number of systems, both fluid and magnet. For example, the gases Ne, Ar, Kr, Xe, N_2, O_2, CO, and CH_4, analyzed by Guggenheim (1945), all give $\beta \approx 0.33 \pm 0.01$. In general, the deviations from the numbers in Table 4.2 do not vary by much more than a few percent from one system to another, either magnet or fluid. (There are, however, some exceptions; e.g., $CrBr_3$ has $\beta \approx 0.37$.) Readers interested in greater experimental detail are advised to consult Stanley (B1971) or the review articles by Heller (1967) and Kadanoff (1967).

We have written down experimental numbers merely to point out two facts:

1. *The predictions of the classical theories are not borne out by experiment.*
2. *The critical exponents are apparently (almost) universal; i.e., they do not seem to depend to any great extent on the particular system, either magnetic or fluid.*

In view of the Lebowitz–Penrose theorem, which states that the classical theories are only valid in the limit of an infinitely weak long-range interaction, fact 1 may be considered as understood. Realistic systems no doubt have finite-range interactions, but these may be effectively long range. For example, in three dimensions each atom in a face-centered-cubic structure has 12 nearest neighbors, so in this case, even with nearest-neighbor interactions, the "range of interaction" is rather long. This in a way explains why the classical theories work well away from the critical point. Presumably one has to get within roughly the inverse range of interaction of the critical point to see deviations from classical behavior.

The second fact is very intriguing and is theoretically a very open question at the moment. Presumably it means that there are only a few parameters, such as dimensionality and underlying "symmetries" which are relevant near the critical point, but what these parameters are nobody knows.

The best we can do at the moment is to examine, in detail, various models that can be solved exactly in the hope that we shall gain some insight into the intricacies of the transition region. In this respect the most studied model to date has been the Ising model, which has been solved exactly in two dimensions but not in three dimensions. Various systematic numerical methods in three dimensions, however, lead to estimates of exponents which are not too different from the experimental values (at least they are closer to the experimental values than the classical theories), so the hope is that the Ising model will provide us with the necessary insight to understand more complicated systems. With this in mind we present in the next two chapters a rather full account of the Ising model.

Some recent heuristic developments in the study of critical exponents have come from attempts to find relations between various exponents. Fisher and Essam [Essam and Fisher (1963a)] were the first to suggest the following relation:

$$\alpha' + 2\beta + \gamma' = 2 \tag{6.5}$$

on the basis of a particular (droplet) model of condensation. Immediately afterward Rushbrooke (1963) pointed out that from thermodynamics alone the inequality

$$\alpha' + 2\beta + \gamma' \geq 2 \tag{6.6}$$

must be satisfied. By now there are a multitude of exponents, some measurable and others not, and corresponding inequalities between them. The reader is referred to Stanley (B1971) for details. In the following section we derive Equation 6.6 and the inequality

$$\alpha' + \beta(1 + \delta) \geq 2, \tag{6.7}$$

which is due to Griffiths (1965a).

4-7 Critical-exponent Inequalities

We shall consider only magnetic systems and prove first the Rushbrooke inequality

$$\alpha' + 2\beta + \gamma' \geq 2, \tag{7.1}$$

where from Table 4.1 α', β, and γ' are the low-temperature exponents for the zero-field specific heat, the spontaneous magnetization, and the zero-field isothermal susceptibility, respectively.

The inequality 7.1 is an immediate consequence of the thermodynamic identity

$$C_H - C_M = T\frac{(\partial M/\partial T)_H^2}{(\partial M/\partial H)_T}, \tag{7.2}$$

where C_H is the specific heat at constant field H and C_M is the specific heat at constant magnetization M. This relation is analogous to the fluid relation, Equation 6.15 of Chapter 2, connecting C_P, C_V, α, and K_T.

As we have noted, thermodynamic relations for magnetic systems are obtained simply from corresponding relations for fluid systems by replacing V by M and P by $-H$. For completeness, however, we give here a brief derivation of Equation 7.2.

By definition

$$C_H = T\left(\frac{\partial S}{\partial T}\right)_H \quad \text{and} \quad C_M = T\left(\frac{\partial S}{\partial T}\right)_M, \tag{7.3}$$

where S is the entropy. In terms of the Helmholtz free energy Ψ

$$S = -\left(\frac{\partial \Psi}{\partial T}\right)_H \quad \text{and} \quad M = -\left(\frac{\partial \Psi}{\partial H}\right)_T. \tag{7.4}$$

Now

$$\left(\frac{\partial S}{\partial T}\right)_M = \left(\frac{\partial S}{\partial T}\right)_H + \left(\frac{\partial S}{\partial H}\right)_T\left(\frac{\partial H}{\partial T}\right)_M. \tag{7.5}$$

In addition, we have the Maxwell relation (see Problem 2 of Chapter 2)

$$\left(\frac{\partial S}{\partial H}\right)_T = -\frac{\partial^2 \Psi}{\partial T\, \partial H} = \left(\frac{\partial M}{\partial T}\right)_H \tag{7.6}$$

and the chain relation (see Problem 5 of Chapter 2)

$$\left(\frac{\partial H}{\partial T}\right)_M \left(\frac{\partial T}{\partial M}\right)_H \left(\frac{\partial M}{\partial H}\right)_T = -1. \tag{7.7}$$

From the definitions 7.3 and Equation 7.5 we have

$$C_M = C_H + T\left(\frac{\partial S}{\partial H}\right)_T \left(\frac{\partial H}{\partial T}\right)_M \tag{7.8}$$

and from Equations 7.6 and 7.7 the second term on the right-hand side of Equation 7.8 is $-T(\partial M/\partial T)_H^2/(\partial M/\partial H)_T$, which gives the relation 7.2.

Since specific heat is essentially a mean-square energy fluctuation (see Problem 4 of Chapter 3) it follows that C_M is nonnegative. This also follows from the fact that the free energy is a convex function of temperature at fixed magnetization. The relation 7.2, therefore, gives

$$C_H \geq T\frac{(\partial M/\partial T)_H^2}{(\partial M/\partial H)_T}. \tag{7.9}$$

In the limit of zero field $M = M_0$ is the spontaneous magnetization and $(\partial M/\partial H)_{H=0} = \chi_0$ is the zero-field isothermal susceptibility, so Equation 7.9 gives

$$C_{H=0} \geq \frac{T(\partial M_0/\partial T)^2}{\chi_0}. \tag{7.10}$$

Taking logarithms of both sides we obtain

$$\log C_{H=0} \geq 2 \log\left(\left|\frac{\partial M_0}{\partial T}\right|\right) - \log \chi_0 + \log T.$$

Dividing by $\log|T_c - T|$ and taking the limit $T \to T_c-$ then gives, from the definitions of the exponents α' β, and γ' (see Table 4.1),

$$-\alpha' \leq 2(\beta - 1) + \gamma', \tag{7.11}$$

which is the inequality 7.1 rearranged.

We now derive the Griffiths inequality

$$\alpha' + \beta(1 + \delta) \geq 2, \tag{7.12}$$

where α' and β are as above and δ is the exponent characterizing the magnetic isotherm at $T = T_c$ and $H \to 0$, i.e.,

$$H \sim |M|^{\delta} \operatorname{sgn}(M) \qquad \text{at } T = T_c \text{ as } H \to 0. \tag{7.13}$$

This inequality is also a consequence of thermodynamics [Rushbrooke (1965)], but we present here Griffiths' (1965a) original derivation based on convexity arguments. The Rushbrooke inequality 7.1 can be derived in a similar manner [Fisher (1967)].

Taking magnetization M and temperature T as independent variables in the free energy Ψ, we have

$$H = \left(\frac{\partial \Psi}{\partial M}\right)_T \qquad \text{and} \qquad S = -\left(\frac{\partial \Psi}{\partial T}\right)_M, \tag{7.14}$$

where H is the magnetic field and S the entropy. For $T < T_c$ and M less than the spontaneous magnetization M_0, H vanishes, by definition 7.14. It follows from Equation 7.14 that

$$\Psi(T, M) = \Psi(T, 0) \qquad \text{when} \qquad M \leq M_0(T). \tag{7.15}$$

Let now M_1 be the spontaneous magnetization for some $T_1 < T_c$. Since $\Psi(T, M)$ is a convex function of T for fixed magnetization we have

$$\left(\frac{\partial \Psi}{\partial T}\right)_{T=T_1} \geq [\Psi(T_2, M) - \Psi(T_1, M)](T_2 - T_1)^{-1} \tag{7.16}$$

whenever $T_1 < T_2 \leq T_c$. Geometrically this means that for a convex function the tangent to the curve always lies above the curve, as shown in Figure 4.14 (see Problem 4 of Chapter 2). Since $S = -\partial\Psi/\partial T$ we deduce from Equation 7.16 with $T_2 = T_c$ that

$$\begin{aligned} \Psi(T_c, M_1) &\leq \Psi(T_1, M_1) + (T_1 - T_c)S(T_1, M_1) \\ &= \Psi(T_1) + (T_1 - T_c)S(T_1). \end{aligned} \tag{7.17}$$

Similarly,

$$\Psi(T_1) \leq \Psi(T_c, 0) + (T_c - T_1)S(T_c). \tag{7.18}$$

From Equations 7.17 and 7.18 it follows that

$$\Psi(T_c, M_1) - \Psi(T_c, 0) \leq (T_c - T_1)[S(T_c) - S(T_1)]. \tag{7.19}$$

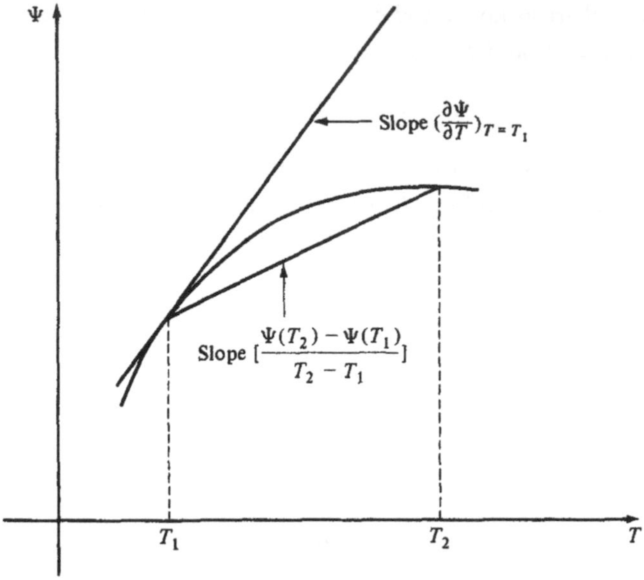

FIGURE 4.14. Free energy Ψ as a function of T for fixed volume (see the text).

Now when $T_1 \to T_c-$ the left-hand side of Equation 7.19, by definition (Equation 7.14 and the definition of exponent β), behaves as (see Problem 8)

$$M_1^{1+\delta} \sim (T_c - T_1)^{\beta(1+\delta)} \tag{7.20}$$

and the right-hand side behaves as

$$(T_c - T_1)(T_c - T_1)^{1-\alpha'} \tag{7.21}$$

since $C_{H=0} = (\partial S/\partial T)_{H=0} \sim (T_c - T_1)^{-\alpha'}$ as $T_1 \to T_c-$. We therefore have a contradiction unless

$$\beta(1 + \delta) \geq 2 - \alpha'$$

which is Griffiths' inequality (7.12).

A number of heuristic arguments known generally as scaling and homogeneity [see Fisher (1967) and Stanley (B1971)] suggest that the Rushbrooke and Griffiths inequalities (and others) may in fact be equalities, i.e., that

$$\alpha' + 2\beta + \gamma' = 2$$

and (7.22)

$$\alpha' + \beta(1 + \delta) = 2.$$

Both these relations are satisfied by the classical exponents ($\alpha' = 0$, $\beta = \frac{1}{2}$, $\gamma' = 1$, and $\delta = 3$) and they also seem to be satisfied by the two-dimensional Ising-model exponents (only α' and β are known exactly, however). The numerical estimates also appear to satisfy Equation 7.22, so it would certainly be worthwhile to give some rigorous argument for these conjectures. We will return to this point again in Chapter 6.

PROBLEMS

1. The Laplace transform $\mathscr{L}\{h\}$ of a function $h(x)$ is defined by

$$\mathscr{L}\{h\} = \int_0^\infty e^{-sx} h(x)\, dx.$$

If

$$F(t) = \int_0^t f(t-x) g(x)\, dx$$

prove that

$$\mathscr{L}\{F\} = \mathscr{L}\{f\}\mathscr{L}\{g\}.$$

(This is the Laplace convolution formula.)

2. The partition function $Z(l)[l = L - (N-1)a]$ for the Takahashi gas (Equation 2.17) is given by

$$\int_0^\infty e^{-sl} Z(l)\, dl = s^{-2}[K(s)]^{N-1}, \tag{1}$$

where

$$K(s) = \int_0^\infty \exp[-sx - \beta v(x)]\, dx.$$

Assuming that the left-hand side of Equation 1 can be evaluated by Laplace's method (for large L and N with L/N fixed), show that

$$\log Z(l) \sim sl + (N-1)\log K(s) - 2\log s,$$

where

$$s = \left[\frac{\partial}{\partial l}\log Z(l)\right]_{l=l} = \beta P(l)$$

and $P(l)$ is the pressure.

3. Consider the potential [after Gallavotti et al. (1968)]

$$\phi(r) = \begin{cases} \infty & |r| < R \\ v(r) & |r| \geq R \end{cases}$$

in one dimension such that $|v(r)| \leq h(r)$ with $h(r)$ and $rh(r)$ decreasing functions of r and

$$I(R) = \int_0^\infty (R + r)h(r)\,dr < \infty.$$

Show that under these conditions, for allowed configurations,

$$\sum_{x_i \leq \xi} \sum_{x_j > \xi} |\phi(x_j - x_i)| \leq R^{-2} I(R);$$

i.e., the attractive interaction energy between particles to the left of some point ξ and particles to the right of ξ is bounded.

4. Consider the model partition function [Uhlenbeck and Ford (B1963)] $Z(V, N)$ defined by (V integral)

$$Z(V, N) = \sum_{k=0}^{V} \frac{V!}{(V - N + k)!(N - k)!}.$$

Show that

(a) the grand canonical partition function is given by

$$Z_G(V, z) = \sum_{N=0}^{\infty} z^N Z(V, N) = \left(\frac{z^{V+1} - 1}{z - 1}\right)(1 + z)^V;$$

(b) the grand-canonical potential $\chi(z)$ is given in the thermodynamic limit by

$$\chi(z) = \lim_{V \to \infty} V^{-1} \log Z_G(V, z) = \begin{cases} \log(1 + z) & \text{when } |z| \leq 1 \\ \log z(1 + z) & \text{when } |z| \geq 1. \end{cases}$$

(c) From the definition of p and v, i.e.,

$$\beta p = \chi(z), \qquad v^{-1} = z \frac{\partial}{\partial z} \chi(z)$$

and the result of part (b) obtain p as a function v. Deduce that

$$\beta p = \log 2 \text{ when } \tfrac{2}{3} \leq v \leq 2.$$

5. Show that the Maxwell construction on the free energy [i.e., replacing $f(v)$ by its convex envelope] is equivalent to the equal-area construction on the isotherms.

6. Deduce the classical critical exponents (see Tables 4.1 and 4.2) $\alpha' = \alpha = 0_{\text{disc}}$, $\gamma' = \gamma = 1$, $\beta = \frac{1}{2}$, and $\delta = 3$ from the van der Waals equation of state.

7. Show from Equation 5.31 that the magnitude of the specific heat-jump discontinuity in the Curie–Weiss theory is $3k/2$.

8. From the definition of the critical exponent λ of a function $f(x)$, i.e.,

$$f(x) \sim x^{\lambda} \text{ as } x \to 0+$$

if

$$\lim_{x \to 0+} \frac{\log f(x)}{\log x} = \lambda \text{ exists,}$$

show that
(a) if $f(x) \sim x^{\lambda}$ and $g(x) \sim x^{\delta}$ as $x \to 0+$ and $f(x) < g(x)$ for sufficiently small positive x, then $\lambda > \delta$;
(b) if $f(x) \sim x^{\lambda}$ as $x \to 0+$ and $\lambda > -1$, then $\int_0^x f(t)\, dt \sim x^{\lambda+1}$ as $x \to 0+$;
(c) if $f(x) \sim x^{\lambda}$ and $g(x) \sim x^{\delta}$ as $x \to 0+$, then $f(g(x)) \sim x^{\lambda\delta}$ as $x \to 0+$.

CHAPTER 5

The Ising Model: Algebraic Approach

5-1 History and Formulation

The Ising model is perhaps the simplest system that undergoes a nontrivial phase transition, and for reasons that will become clear, now occupies an almost unique place in theoretical physics. The literature on the Ising model is enormous; since the original contribution of Ising (1925) almost 500 research papers have been published on the subject.

The 1925 paper of Ising was essentially his Ph.D. thesis. He succeeded in solving the one-dimensional model exactly and found to nobody's surprise these days that there was no phase transition. In this paper Ising gives credit to his supervisor Wilhelm Lenz for inventing the model, but curiously enough Lenz has never been associated with the model since.

The model was originally proposed as a model for ferromagnetism, the "spins" being scalar spins capable of only two orientations, "up" or "down," as discussed in Section 4-5. It was a great disappointment, especially to Ising, that the model did not exhibit ferromagnetism in one dimension. Ising, in fact, gave arguments why the model should not exhibit ferromagnetism in two and three dimensions (we know now that this is not the case), so it was natural that the model was subsequently neglected. It was taken up again in the 1930s by Bragg and Williams (1934, 1935), Bethe (1935), Peierls

(1936a, 1936b), and others, but mainly as a model for a binary alloy (spin up corresponding to an atom of type A and spin down to an atom of type B) rather than of a ferromagnet.

Peierls (1936a) gave a proof of the existence of ferromagnetism for the two-dimensional model which was unfortunately incorrect. The error was only noticed recently (by Fisher and Sherman) and the proof has since been made completely rigorous by Griffiths (1964).

The major breakthrough came from Kramers and Wannier (1941), who formulated the model as a matrix problem and from symmetry considerations actually succeeded in locating the critical point exactly for the two-dimensional model. They did not, however, solve the problem completely; this was left to Onsager (1944), who obtained the complete solution (i.e., the partition function) for the two-dimensional model, in the absence of an external magnetic field. The Onsager solution was really the first nontrivial demonstration of the existence of a phase transition from the partition function alone.

After Onsager, papers began to flow forth. Since an excellent review has recently been published [Brush (1967)], we shall not dwell further on the history at this stage. We shall, however, have a little more to say on the historical side as we progress, but first let us define the model precisely.

Consider a lattice (e.g., the two-dimensional square lattice shown in Figure 5.1) with spins μ_P capable of two values, $+1$ (spin up) and -1 (spin down), located on the vertices P of the lattice. If there are N vertices, or lattice sites, there are a total of 2^N possible configurations of spins (two for each site) on the lattice, a configuration $\{\mu\}$ being specified by the N spin variables, i.e.,

$$\{\mu\} = (\mu_P\colon P \text{ a lattice point}).$$

In a configuration $\{\mu\}$ the interaction energy is defined by

$$E\{\mu\} = -J \sum_{P,Q}{}^{*} \mu_P \mu_Q - H \sum_P \mu_P, \tag{1.1}$$

where the starred sum is over nearest-neighbor lattice points P and Q. J is the "coupling constant" and H is the external magnetic field. The first term in Equation 1.1 represents the interaction energy between spins and the second term the interaction between the individual spins and the external field.

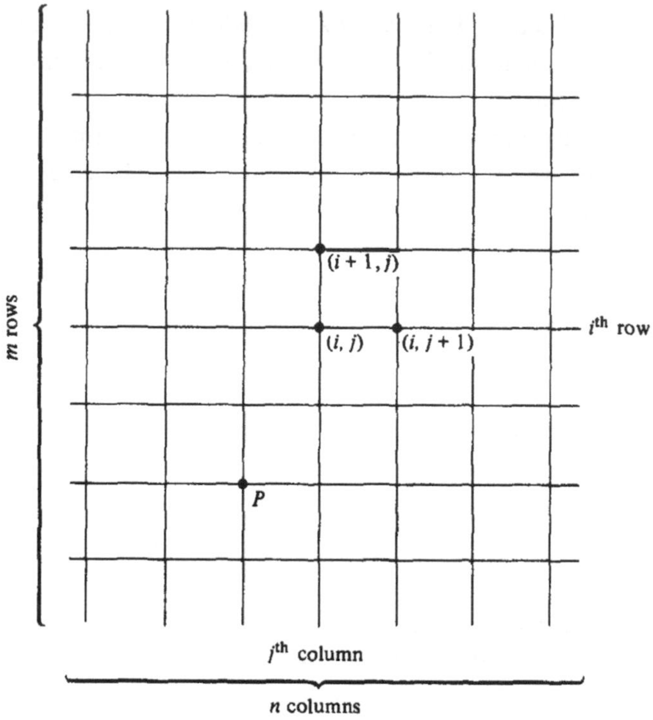

FIGURE 5.1. Two-dimensional square Ising lattice. Spins occupy the vertices P of the lattice.

For example, the interaction energy for the two-dimensional lattice shown in Figure 5.1 with lattice points denoted by (i, j) is given by

$$E\{\mu\} = -J\left(\sum_{i=1}^{m-1}\sum_{j=1}^{n}\mu_{i,j}\mu_{i+1,j} + \sum_{i=1}^{m}\sum_{j=1}^{n-1}\mu_{i,j}\mu_{i,j+1}\right) - H\sum_{i=1}^{m}\sum_{j=1}^{n}\mu_{i,j}. \quad (1.2)$$

In one dimension the lattice becomes a chain of N spins, as shown in Figure 5.2, and

$$E\{\mu\} = -J\sum_{i=1}^{N-1}\mu_i\mu_{i+1} - H\sum_{i=1}^{N}\mu_i. \quad (1.3)$$

In most cases it is convenient to consider periodic lattices instead of the open lattices above (Equations 1.2 and 1.3). For example, in one dimension

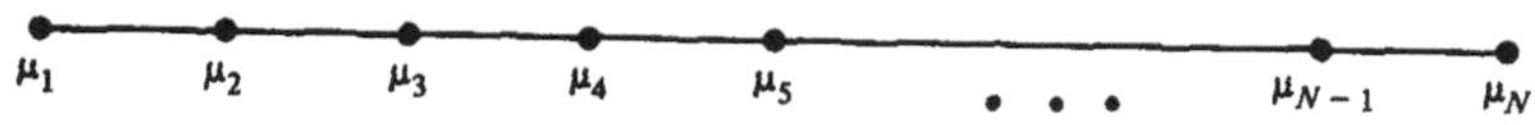

FIGURE 5.2. One-dimensional Ising chain.

we consider a ring of spins, as shown in Figure 5.3, with spin N coupled to spin 1. The interaction energy is then given by Equation 1.3 with $N-1$ replaced by N in the first sum, with the convention that $\mu_{N+1} = \mu_1$. In two dimensions we achieve periodicity by wrapping the lattice on a *torus* so that

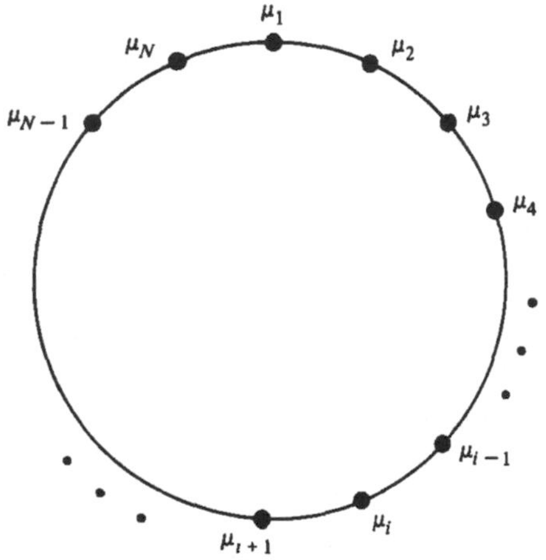

FIGURE 5.3. Closed Ising chain with spin N coupled to spin 1.

the nth column is coupled to the first column and the mth row is coupled to the first row. The interaction energy is given by Equation 1.2 with $m-1$ and $n-1$ replaced by m and n, respectively, with the convention that

$$\begin{aligned} \mu_{m+1,j} &= \mu_{1,j} \qquad j = 1, 2, \ldots, n, \\ \mu_{i,n+1} &= \mu_{i,1}, \qquad i = 1, 2, \ldots, m. \end{aligned} \tag{1.4}$$

The generalization to higher-dimensional lattices is straightforward.

Notice that the interaction energy between two nearest-neighbor spins P and Q is

$$-J\mu_P\mu_Q = \begin{cases} -J & \text{if both spins are up or both spins are down} \\ +J & \text{if one spin is up and the other spin is down.} \end{cases} \tag{1.5}$$

So if $J > 0$, we get a lower energy for a parallel configuration. The "ground-state energy" (i.e., the minimum $E\{\mu\}$) in this case is, therefore, from Equation 1.1,

$$E_0 = \min_{\{\mu\}} E\{\mu\} = -\tfrac{1}{2}JNq - N|H| \qquad \text{for periodic lattices} \tag{1.6}$$

(for open lattices we have boundary terms), where q is the *coordination number* of the lattice (i.e., the number of nearest-neighbor sites of a given site), so $\sum^{*}_{P,Q} = Nq/2$. The minimum energy (1.6) (when $J > 0$) is achieved when all spins are up (down) for $H > 0 (H < 0)$, and in the absence of an external field ($H = 0$) the minimum is achieved either when all spins are up or all spins are down. For example, the square lattice has $q = 4$ and it is to be noted that for the *open lattice* (Equation 1.2) N in the first term of Equation 1.6 is to be replaced by $N - [(n+m)/2]$ (where $N = mn$). The boundary term $-(n+m)/2$ is, of course, negligible compared with N for large m and n.

Since for $J > 0$ the ground state is achieved when all spins are parallel, we say that $J > 0$, *corresponds to ferromagnetism*. Similarly $J < 0$ *corresponds to antiferromagnetism* since from Equation 1.5 we get a lower energy for an antiparallel configuration. The antiferromagnetic ground state is not always easy to find, however, as the reader may easily convince himself by considering a few examples. [For the one-dimensional chain and the two-dimensional square lattice it is easy, but what about the triangular lattice in two dimensions? (See Problem 1.)]

We have now defined the model. The statistical-mechanical problem is to evaluate the partition function defined by

$$Z_N = \sum_{\{\mu\}} \exp(-\beta E\{\mu\}), \tag{1.7}$$

where the energy $E\{\mu\}$ is given by Equation 1.1. The sum over $\{\mu\}$ denotes a sum over all possible configurations $\mu_P = \pm 1$ of the lattice. For example,

$$Z_N = \sum_{\{\mu\}} \exp\left(\beta J \sum_{i=1}^{N-1} \mu_i \mu_{i+1} + \beta H \sum_{i=1}^{N} \mu_i\right) \tag{1.8}$$

for the one-dimensional open chain (Equation 1.3).

As we will see in Section 5-3, the evaluation of Z_N (Equation 1.8) is trivial, especially when $H = 0$. For the two-dimensional model the problem is extremely complicated and has only been solved in the zero-field case (see Appendix D). The three-dimensional problem is unsolved even for $H = 0$!

In the next section we give two different interpretations of the Ising model, the first as a lattice gas, which in view of our discussion in previous chapters will contain more familiar terminology. As was pointed out in Chapter 4, however, the magnetic interpretation is somewhat less cumbersome, so most of our subsequent discussion will be devoted to it. The second interpretation as a binary alloy is included only for historical interest.

5-2 Lattice Gas and Binary Alloy

The term lattice gas was first coined by Yang and Lee (1952b) and Lee and Yang (1952), although the interpretation of the model as a gas was, in fact, known earlier.

The definition is as follows: Consider a lattice of V sites (V = "volume") and a collection of N particles. The particles are placed on the vertices of the lattice such that not more than one particle can occupy a given lattice site, and only particles on nearest-neighbor lattice sites interact. The interaction potential between two lattice sites P and Q is, therefore,

$$\phi(P, Q) = \begin{cases} \infty & \text{if } P \text{ and } Q \text{ are occupied and } P \equiv Q \text{ (the "hard core")} \\ -A & \text{if } P \text{ and } Q \text{ are nearest-neighbor occupied sites} \\ 0 & \text{otherwise.} \end{cases} \tag{2.1}$$

If we now introduce parameters t_P for each lattice site P of the lattice such that

$$t_P = \begin{cases} 1 & \text{if site } P \text{ is occupied} \\ 0 & \text{if site } P \text{ is unoccupied,} \end{cases} \tag{2.2}$$

the total interaction energy in a given configuration

$$\{t\} = (t_P\colon P \text{ a lattice site})$$

is given by (see Equation 1.1)

$$E\{t\} = -A \sum_{P,Q}{}^{*} t_P t_Q, \tag{2.3}$$

where, as before, the starred sum is over nearest-neighbor lattice sites P and Q. The condition that the total number of particles is N can be written as

$$\sum_P t_P = N \tag{2.4}$$

and the canonical partition function is given by

$$Z(N, V, T) = \sum_{\substack{\{t\} \\ \sum_P t_P = N}} \exp(-\beta E\{t\}). \tag{2.5}$$

To remove the restriction 2.4 on the sum over the configurations $\{t\}$(2^V in total), it is convenient to consider the grand-canonical partition function defined by

$$Z_G(z, V, T) = \sum_{N=0}^{\infty} z^N Z(N, V, T). \tag{2.6}$$

Since the maximum number of particles that can be accommodated on the lattice is V, the upper limit on the sum in Equation 2.6 is $N = V$, and from Equations 2.3 and 2.5 it is obvious that

$$Z_G(z, V, T) = \sum_{\{t\}} z^{\sum_P t_P} \exp(\beta A \sum_{P,Q}^{*} t_P t_Q), \tag{2.7}$$

where the configuration sum is now unrestricted. We now show that the grand-canonical partition function for the lattice gas (Equation 2.7) is equivalent to the canonical partition function (Equation 1.7) for the Ising magnet. To do this we define the variables μ_P by

$$\mu_P = 1 - 2t_P. \tag{2.8}$$

Since t_P is either 0 or 1 the variables μ_P are either $+1(t_P = 0)$ or $-1(t_P = 1)$. Substituting Equation 2.8 into Equation 2.7 shows after an elementary calculation that

$$Z_G(z, V, T) = \exp(\tfrac{1}{2}V \log z + \tfrac{1}{8}\beta VqA) \sum_{\{\mu\}} \exp(\beta J \sum_{P,Q}^{*} \mu_P \mu_Q + \beta H \sum_P \mu_P), \tag{2.9}$$

where

$$\begin{aligned} 4J &= A, \\ \beta H &= -\tfrac{1}{2} \log z - \tfrac{1}{4}\beta qA, \end{aligned} \tag{2.10}$$

and we have used the facts that (see Equation 1.6)

$$\sum_{P,Q}^{*} = \frac{Vq}{2} \tag{2.11}$$

and

$$\sum_{P,Q}^{*} \mu_P = \tfrac{1}{2}q \sum_P \mu_P \tag{2.12}$$

for *periodic lattices* (or more generally if we ignore boundary terms which are in any case negligible in the limit $V \to \infty$). The sum over configurations $\{\mu\}$ in Equation 2.9 is just the canonical partition function for the Ising magnet (Equation 1.7), so it is now straightforward to set up a correspondence between the thermodynamic quantities for the gas and the magnet.

Thus for the lattice gas the *pressure p* from Equation 2.9 is given by

$$\begin{aligned}\beta p &= V^{-1} \log Z_G(z, V, T)\\ &= -\beta\psi - \frac{\beta q J}{2} - \beta H,\end{aligned} \tag{2.13}$$

where ψ is the Ising magnet (canonical) free energy per spin, i.e.,

$$-\beta\psi = V^{-1} \log Z(V, T) \tag{2.14}$$

and $Z(V, T)$ is the canonical partition function Equation 1.7 with V in place of N.

The density ρ, defined by

$$\rho = z\frac{\partial}{\partial z} [V^{-1} \log Z_G(z, V, T)] \tag{2.15}$$

is likewise given in terms of the *magnetization m* from Equations 2.8 and 2.9 by

$$\rho = [Z_G(z, V, T)]^{-1} \sum_{\{t\}} (V^{-1} \sum_P t_P) z^{\sum_P t_P} \exp(\beta A \sum_{P,Q}^{*} t_P t_Q),$$

i.e.,

$$\rho = \tfrac{1}{2}(1 - m), \tag{2.16}$$

where m is defined by

$$\begin{aligned} m &= \frac{\partial}{\partial(\beta H)} [V^{-1} \log Z(V, T)]\\ &= [Z(V, T)]^{-1} \sum_{\{\mu\}} (V^{-1} \sum_P \mu_P) \exp(\beta J \sum_{P,Q}^{*} \mu_P \mu_Q + \beta H \sum_P \mu_P).\end{aligned} \tag{2.17}$$

Similarly, the *isothermal compressibility* K_T defined by

$$K_T = -v^{-1}\left(\frac{\partial v}{\partial p}\right)_T = \rho^{-1}\left(\frac{\partial \rho}{\partial p}\right)_T \tag{2.18}$$

is related to the isothermal susceptibility

$$\chi(T) = \left(\frac{\partial m}{\partial H}\right)_T \tag{2.19}$$

by (see Problem 2)

$$4\rho^2 K_T = \chi(T) \tag{2.20}$$

and for the specific heats

$$\rho C_V = C_M. \tag{2.21}$$

Thermodynamically, as we have mentioned, there is a relation between magnetic and gas–liquid systems. As we have just seen, this relationship takes the form of precise identities for the lattice gas and Ising magnet, but it goes without saying that these identities are not generally valid.

We conclude this section with a brief mention of the binary-alloy interpretation of the Ising model.

For a binary alloy we take the parameters t_P for the lattice gas to be

$$t_P = \begin{cases} 1 & \text{if site } P \text{ is occupied by an A atom} \\ 0 & \text{if site } P \text{ is occupied by a B atom.} \end{cases} \tag{2.22}$$

Then if ε_{AA}, ε_{BB}, and ε_{AB} denote the coupling (interaction) constants for nearest-neighbor A atoms, nearest-neighbor B atoms, and nearest-neighbor A and B atoms, respectively, the interaction energy is given by

$$E\{t\} = -\sum_{P,Q}^{*} \{\varepsilon_{AA} t_P t_Q + \varepsilon_{BB}(1 - t_P)(1 - t_Q) + \varepsilon_{AB}[t_P(1 - t_Q) + t_Q(1 - t_P)]\} \tag{2.23}$$

for a particular configuration $\{t\}$ of A and B atoms. It is left as an exercise for the reader to set up an isomorphism between the binary alloy and the magnet. It is almost obvious, for example, that the binary alloy with the same number of A and B atoms corresponds to a magnet with zero external field.

From this point on we shall mean the Ising magnet when we refer to the Ising model (or problem).

5-3 One-dimensional Model and Transfer Matrix

Let us consider first the one-dimensional open-chain model (Equation 1.3) with zero external field. The problem is to evaluate the partition function (Equation 1.8)

$$Z_N = \sum_{\mu_1 = \pm 1, \ldots, \mu_N = \pm 1} \exp\left(\nu \sum_{i=1}^{N-1} \mu_i \mu_{i+1}\right) \tag{3.1}$$

where $\nu = \beta J$. Because of the one-dimensional structure (cf. the Tonks gas) we can separate off and sum over the Nth spin μ_N to get

$$\begin{aligned} Z_N &= 2 \cosh \nu \sum_{\mu_1 = \pm 1, \ldots, \mu_{N-1} = \pm 1} \exp\left(\nu \sum_{i=1}^{N-2} \mu_i \mu_{i+1}\right) \\ &= (2 \cosh \nu) Z_{N-1}, \end{aligned} \tag{3.2}$$

where we have used the fact that since $\mu_{N-1} = \pm 1$,

$$\begin{aligned}\sum_{\mu_N = \pm 1} \exp(\nu\mu_{N-1}\mu_N) &= \exp(\nu\mu_{N-1}) + \exp(-\nu\mu_{N-1}) \\ &= \exp(\nu) + \exp(-\nu) \\ &= 2\cosh\nu\end{aligned} \tag{3.3}$$

for either $\mu_{N-1} = +1$ or -1. The recurrence relation (3.2) together with the obvious fact that

$$\begin{aligned}Z_2 &= \sum_{\mu_1 = \pm 1, \mu_2 = \pm 1} \exp(\nu\mu_1\mu_2) \\ &= \sum_{\mu_1 = \pm 1} [\exp(\nu\mu_1) + \exp(-\nu\mu_1)] \\ &= 4\cosh\nu\end{aligned} \tag{3.4}$$

gives

$$Z_N = 2(2\cosh\nu)^{N-1}. \tag{3.5}$$

The free energy per spin ψ in the thermodynamic limit is given by

$$\begin{aligned}-\frac{\psi}{kT} &= \lim_{N\to\infty} N^{-1}\log Z_N \\ &= \log(2\cosh\nu),\end{aligned} \tag{3.6}$$

which is a completely analytic function of ν, and hence temperature ($\nu = J/kT$), for all positive temperatures. Hence, as expected, there is no phase transition.

It is obvious that the elementary device leading to the recurrence relation (3.2) will not work if there is an external magnetic field, or in dimensions greater than one even with $H = 0$. Although it is possible to solve the open-chain problem in an external field, it is considerably easier to consider the (closed) periodic chain (Figure 5.3). In the limit $N \to \infty$ we would expect identical results, and this, as we will see, is the case.

For the periodic chain ($\mu_{N+1} = \mu_1$) in an external magnetic field H the problem is to evaluate

$$Z_N = \sum_{\{\mu\}} \exp\left(\nu\sum_{i=1}^{N}\mu_i\mu_{i+1} + B\sum_{i=1}^{N}\mu_i\right), \tag{3.7}$$

where $B = \beta H$. If we define

$$L(\mu_i, \mu_{i+1}) = \exp\left[\nu\mu_i\mu_{i+1} + \frac{B}{2}(\mu_i + \mu_{i+1})\right] \tag{3.8}$$

we can write

$$Z_N = \sum_{\{\mu\}} L(\mu_1, \mu_2)L(\mu_2, \mu_3)\cdots L(\mu_{N-1}, \mu_N)L(\mu_N, \mu_1), \tag{3.9}$$

which has the form of a matrix product. Indeed, if $\mathbf{L}$ is the 2 by 2 *transfer matrix* with elements $L(\mu, \mu')$ defined by Equation 3.8, i.e.,

$$\begin{aligned} \mathbf{L} &= \begin{pmatrix} L(+1, +1) & L(+1, -1) \\ L(-1, +1) & L(-1, -1) \end{pmatrix} \\ &= \begin{pmatrix} \exp(\nu + B) & \exp(-\nu) \\ \exp(-\nu) & \exp(\nu - B) \end{pmatrix}, \end{aligned} \tag{3.10}$$

we have, after summing in Equation 3.9 over $\mu_2 = \pm 1, \ldots, \mu_N = \pm 1$, that

$$Z_N = \sum_{\mu_1 = \pm 1} L^N(\mu_1, \mu_1), \tag{3.11}$$

where $L^N(\mu, \mu')$ denotes the (μ, μ') elements of the matrix $\mathbf{L}$ raised to the Nth power. Z_N (Equation 3.11) is therefore the sum of the diagonal elements of $\mathbf{L}^N$, i.e., the trace of $\mathbf{L}^N$. Now since the trace of a matrix is the sum of its eigenvalues and the eigenvalues of $\mathbf{L}^N$ are the eigenvalues of $\mathbf{L}$ raised to the Nth power, we have that

$$\begin{aligned} Z_N &= \mathrm{Tr}(\mathbf{L}^N) \\ &= \lambda_1^N + \lambda_2^N, \end{aligned} \tag{3.12}$$

where λ_1 and λ_2 are the eigenvalues of the matrix $\mathbf{L}$ (Equation 3.10). The eigenvalue equation is

$$\mathrm{Det}\begin{pmatrix} \exp(\nu + B) - \lambda & \exp(-\nu) \\ \exp(-\nu) & \exp(\nu - B) - \lambda \end{pmatrix} = \lambda^2 - 2\lambda e^{\nu} \cosh B + 2 \sinh 2\nu = 0, \tag{3.13}$$

which gives

$$\left.\begin{matrix} \lambda_1 \\ \lambda_2 \end{matrix}\right\} = e^{\nu} \cosh B \pm (e^{2\nu} \sinh^2 B + e^{-2\nu})^{1/2} \tag{3.14}$$

for the eigenvalues of $\mathbf{L}$.

Noting that λ_2/λ_1 is strictly less than unity for all $v > 0$ we have that

$$\begin{aligned}
-\frac{\psi}{kT} &= \lim_{N\to\infty} N^{-1} \log Z_N \\
&= \lim_{N\to\infty} N^{-1} \log\{\lambda_1^N [1 + (\lambda_2/\lambda_1)^N]\} \\
&= \log \lambda_1 + \lim_{N\to\infty} N^{-1} \log[1 + (\lambda_2/\lambda_1)^N] \\
&= \log \lambda_1 \\
&= \log[e^v \cosh B + (e^{2v} \sinh^2 B + e^{-2v})^{1/2}].
\end{aligned} \tag{3.15}$$

When $B = 0$ the right-hand side of Equation 3.15 is $\log(2 \cosh v)$, which is precisely the expression (Equation 3.6) for the open chain. Further, from Equation 3.14 with $B = 0$,

$$\begin{aligned}
\lambda_1 &= 2 \cosh v, \\
\lambda_2 &= 2 \sinh v,
\end{aligned} \tag{3.16}$$

so, from Equation 3.12, the partition function for the finite chain is given by

$$Z_N = (2 \cosh v)^N + (2 \sinh v)^N, \tag{3.17}$$

which obviously differs from the open-chain-result, Equation 3.5. In the thermodynamic limit, however, only the maximum eigenvalue (λ_1) contributes. We will see that this is generally the case in applications of the transfer-matrix method.

Note that the magnetization per spin computed from Equation 3.15 is is given by

$$\begin{aligned}
m &= \frac{\partial}{\partial B}\left(-\frac{\psi}{kT}\right) \\
&= \sinh B(\sinh^2 B + e^{-4v})^{-1/2}.
\end{aligned} \tag{3.18}$$

In zero field ($B = 0$) the (spontaneous) magnetization is zero, as expected, for all finite temperatures ($v > 0$).

5-4 Transfer Matrix for the Two- and Higher-dimensional Models

The advantage of the matrix method is that it can be easily generalized to two or more dimensions. Let us consider first the two-dimensional problem on a square lattice wrapped on a cylinder (i.e., periodic in columns but not rows), as shown in Figure 5.4. The interaction energy is given by (Equation 1.2 with $\mu_{i,n+1} = \mu_{i,1}$)

$$E\{\mu\} = -J\sum_{i=1}^{m-1}\sum_{j=1}^{n}\mu_{i,j}\mu_{i+1,j} - J\sum_{i=1}^{m}\sum_{j=1}^{n}\mu_{i,j}\mu_{i,j+1} - H\sum_{i=1}^{m}\sum_{j=1}^{n}\mu_{i,j}. \tag{4.1}$$

If we now denote a column configuration by σ_j, i.e.,

$$\sigma_j = (\mu_{1,j}, \mu_{2,j}, \ldots, \mu_{m,j}) \tag{4.2}$$

(there are a total of 2^m possible configurations for each column), we can write $E\{\mu\}$ as a sum of two terms—the interaction energy of columns and the interaction energy between nearest-neighbor columns—i.e., if we define

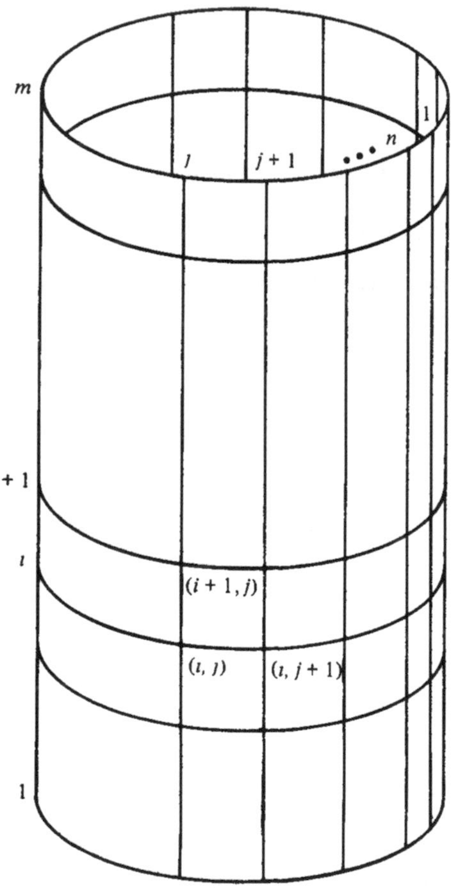

FIGURE 5.4. Two-dimensional Ising lattice wrapped on a cylinder.

$$V_1(\sigma_j) = -J\sum_{i=1}^{m-1}\mu_{i,j}\mu_{i+1,j} - H\sum_{i=1}^{m}\mu_{i,j} \tag{4.3}$$

to be the interaction energy of the jth column, and

$$V_2(\sigma_j, \sigma_{j+1}) = -J\sum_{i=1}^{m}\mu_{i,j}\mu_{i,j+1} \tag{4.4}$$

to be the interaction energy between the jth and $(j+1)$th columns, we can write $(\sigma_{n+1} = \sigma_1)$

$$\begin{aligned} E\{\mu\} &= E\{\sigma_1, \sigma_2, \ldots, \sigma_n\} \\ &= \sum_{j=1}^{n}[V_1(\sigma_j) + V_2(\sigma_j, \sigma_{j+1})]. \end{aligned} \tag{4.5}$$

The partition function can then be written as

$$\begin{aligned} Z_{n,m} &= \sum_{\{\mu\}}\exp(-\beta E\{\mu\}) \\ &= \sum_{\sigma_1,\ldots,\sigma_n}\exp\left[-\beta\left(\sum_{j=1}^{n}\{V_1(\sigma_j) + V_2(\sigma_j,\sigma_{j+1})\}\right)\right] \\ &= \sum_{\sigma_1,\ldots,\sigma_n} L(\sigma_1,\sigma_2)L(\sigma_2,\sigma_3)\cdots L(\sigma_{n-1},\sigma_n)L(\sigma_n,\sigma_1) \\ &= \sum_{\sigma_1} L^n(\sigma_1,\sigma_1), \end{aligned} \tag{4.6}$$

where from Equations 4.3 and 4.4 with $\sigma = (\mu_1, \mu_2, \ldots, \mu_m)$ and $\sigma' = (\mu_1', \mu_2', \ldots, \mu_m')$,

$$\begin{aligned} L(\sigma, \sigma') &= \exp[-\beta V_1(\sigma)]\exp[-\beta V_2(\sigma, \sigma')] \\ &= \exp\left(v\sum_{i=1}^{m-1}\mu_i\mu_{i+1} + B\sum_{i=1}^{m}\mu_i\right)\exp\left(v\sum_{i=1}^{m}\mu_i\mu_i'\right) \end{aligned} \tag{4.7}$$

with

$$v = \beta J \qquad \text{and} \qquad B = \beta H. \tag{4.8}$$

Alternatively, we can take the symmetric matrix (as we did for the one-dimensional model)

$$\begin{aligned} L'(\sigma, \sigma') &= \exp\left[-\frac{\beta}{2}V_1(\sigma)\right]\exp[-\beta V_2(\sigma, \sigma')]\exp\left[-\frac{\beta}{2}V_1(\sigma')\right] \\ &= \exp\left(\frac{v}{2}\sum_{i=1}^{m-1}\mu_i\mu_{i+1} + \frac{B}{2}\sum_{i=1}^{m}\mu_i\right)\exp\left(v\sum_{i=1}^{m}\mu_i\mu_i'\right) \\ &\quad\times\exp\left(\frac{v}{2}\sum_{i=1}^{m-1}\mu_i'\mu_{i+1}' + \frac{B}{2}\sum_{i=1}^{m}\mu_i'\right). \end{aligned} \tag{4.9}$$

The final result is, of course, the same. $L^n(\sigma, \sigma)$ in Equation 4.6 denotes the (σ, σ) component of the 2^m by 2^m matrix $\mathbf{L}$ with elements 4.7, raised to the nth power, i.e., by analogy with Equation 3.12,

$$
\begin{aligned}
Z_{n,m} &= \mathrm{Tr}(\mathbf{L}^n) \\
&= \sum_{j=1}^{2^m} \lambda_j^n,
\end{aligned}
\tag{4.10}
$$

where $\lambda_1 > \lambda_2 \geq \cdots \geq \lambda_{2^m}$ are the eigenvalues of the 2^m by 2^m matrix $\mathbf{L}$. If now, in the thermodynamic limit, we allow n to approach infinity before m (this is convenient rather than necessary), we have for the free energy per spin ψ,

$$
\begin{aligned}
-\frac{\psi}{kT} &= \lim_{m\to\infty} \lim_{n\to\infty} (mn)^{-1} \log Z_{n,m} \\
&= \lim_{m\to\infty} m^{-1} \log \lambda_1 + \lim_{m\to\infty} \left[\lim_{n\to\infty} (mn)^{-1} \log\left(1 + \sum_{j=2}^{2^m} (\lambda_j/\lambda_1)^n\right)\right] \\
&= \lim_{m\to\infty} m^{-1} \log \lambda_1 .
\end{aligned}
\tag{4.11}
$$

So the problem has again been reduced to finding the largest eigenvalue of a matrix, but notice the dramatic effect of dimensionality: In one dimension we had only to find the largest eigenvalue of a 2 by 2 matrix, but in two dimensions we have to find the largest eigenvalue of a 2^m by 2^m matrix and then let m approach infinity!

The manipulation leading to Equation 4.11 can be repeated essentially word for word in three and higher dimensions. Thus in three dimensions, for example, we define σ_j to be the configuration of the jth (two-dimensional) plane and build the lattice up by planes. $V_1(\sigma_j)$ then represents the interaction energy of a plane and $V_2(\sigma_j, \sigma_{j+1})$ the interaction energy of nearest-neighbor jth and $(j+1)$th planes, and so forth. The interested reader can fill in the details for himself.

The two-dimensional problem with $H = 0$ was solved by Onsager (1944) in one of the most celebrated articles of modern times. The corresponding problem in three dimensions and the two-dimensional problem with $H \neq 0$ are unsolved.

There have been many simplifications in the derivation of the Onsager result since 1944, but even the simplest are rather complicated. We shall present in the following section only a statement and discussion of Onsager's result. A detailed derivation based on Equation 4.11 is given in Appendix D.

5-5 The Onsager Solution of the Two-dimensional Model

By a masterly application of Lie algebras and group representations, Onsager found the largest eigenvalue of the transfer matrix, Equation 4.7, with $H = 0$, to be

$$\lambda_1 = (2 \sinh 2\nu)^{m/2} \exp[\tfrac{1}{2}(\gamma_1 + \gamma_3 + \cdots + \gamma_{2m-1})], \tag{5.1}$$

where γ_k is defined by

$$\cosh \gamma_k = \cosh 2\nu \coth 2\nu - \cos\left(\frac{\pi k}{m}\right) \tag{5.2}$$

and

$$\nu = \frac{J}{kT}. \tag{5.3}$$

A derivation of this result is given in Appendix D.

The free energy per spin ψ in the thermodynamic limit is then given, from Equations 4.11 and 5.1, by

$$\begin{aligned} -\frac{\psi}{kT} &= \lim_{m\to\infty} m^{-1} \log \lambda_1 \\ &= \tfrac{1}{2}\log(2\sinh 2\nu) + \lim_{m\to\infty} (2m)^{-1} \sum_{k=0}^{m-1} \gamma_{2k+1}. \end{aligned} \tag{5.4}$$

In the limit $m \to \infty$ the sum in Equation 5.4 approaches an integral (see Equation 5.2), so that

$$-\frac{\psi}{kT} = \tfrac{1}{2}\log(2\sinh 2\nu) + (2\pi)^{-1}\int_0^\pi \cosh^{-1}(\cosh 2\nu \coth 2\nu - \cos\theta)\, d\theta. \tag{5.5}$$

Use of the identity

$$\cosh^{-1}|z| = \pi^{-1}\int_0^\pi \log[2(z - \cos\phi)]\, d\phi \tag{5.6}$$

allows us to write

$$-\frac{\psi}{kT} = \frac{1}{2}\log(2\sinh 2\nu) + \frac{1}{2}\log 2 + \frac{1}{2\pi^2}\iint_0^\pi \log(\cosh 2\nu \coth 2\nu - \cos\theta - \cos\phi)\, d\theta\, d\phi, \tag{5.7}$$

which gives the symmetric Onsager formula

$$-\frac{\psi}{kT}=\log 2+\frac{1}{2\pi^2}\iint_0^\pi \log[\cosh^2 2v-\sinh 2v(\cos\theta_1+\cos\theta_2)]\,d\theta_1\,d\theta_2\,. \tag{5.8}$$

From Equation 5.8 the internal energy U is given by

$$\begin{aligned} U &= -kT^2\frac{\partial}{\partial T}\frac{\psi}{kT} \\ &= J\frac{\partial}{\partial v}\frac{\psi}{kT} \\ &= -J\coth 2v \\ &\quad\times\left[1+(\sinh^2 2v-1)\frac{1}{\pi^2}\iint_0^\pi \frac{d\theta_1\,d\theta_2}{\cosh^2 2v-\sinh 2v(\cos\theta_1+\cos\theta_2)}\right]. \end{aligned} \tag{5.9}$$

The integral in Equation 5.9 diverges logarithmically (at the origin $\theta_1=\theta_2=0$) when $\cosh^2 2v=2\sinh 2v$. To see this, note that in the neighborhood of the origin

$$\cos\theta_1+\cos\theta_2\sim 2-\tfrac{1}{2}(\theta_1^2+\theta_2^2)+\cdots,$$

so when $\delta=\cosh^2 2v-2\sinh 2v\sim 0$,

$$\begin{aligned} &\frac{1}{\pi^2}\iint_0^\pi \frac{d\theta_1\,d\theta_2}{\cosh^2 2v-\sinh 2v(\cos\theta_1+\cos\theta_2)} \\ &\qquad\sim\frac{1}{\pi^2}\iint_0 \frac{d\theta_1\,d\theta_2}{\delta+\frac{1}{2}\sinh 2v(\theta_1^2+\theta_2^2)} \\ &\qquad=\frac{2}{\pi}\int_0 \frac{r\,dr}{\delta+\frac{1}{2}\sinh 2v\,r^2} \\ &\qquad\sim-\frac{2}{\pi\sinh 2v}\log|\delta|, \end{aligned} \tag{5.10}$$

where in the last step we have transformed to polar coordinates ($\theta_1^2+\theta_2^2=r^2$, etc.).

There is a singularity, or phase-transition point, therefore, when

$$\delta=\cosh^2 2v-2\sinh 2v=0, \tag{5.11}$$

i.e., when $\nu = \nu_c = J/kT_c$, given by

$$\sinh 2\nu_c = 1. \tag{5.12}$$

In Equation 5.9 for the internal energy U, the integral, Equation 5.10, is multiplied by $(\sinh^2 2\nu - 1)$, which is zero at the critical point $\nu = \nu_c$. It follows that the internal energy is continuous at $\nu = \nu_c$ and that in the neighborhood of ν_c,

$$U \sim -J \coth 2\nu_c[1 + A(\nu - \nu_c) \log|\nu - \nu_c|], \tag{5.13}$$

where A is a constant. From this result it follows that the specific heat, defined by

$$C = \frac{\partial U}{\partial T}, \tag{5.14}$$

has a symmetrical logarithmic divergence (i.e., $C \sim B \log|\nu - \nu_c|$) at the critical point ν_c, as shown in Figure 5.5.

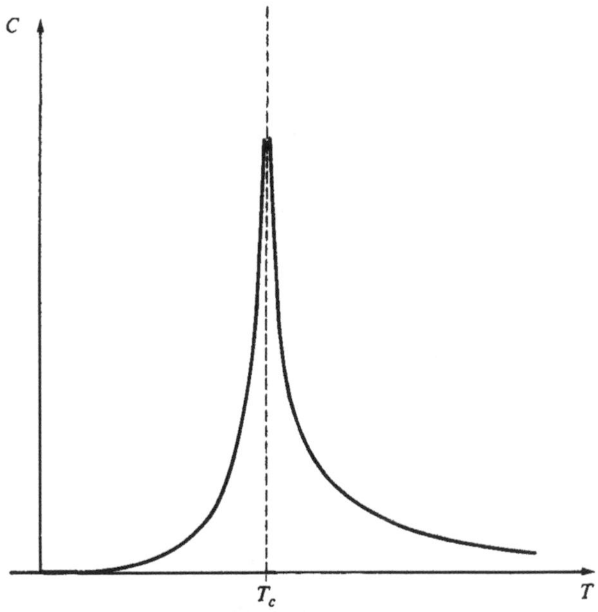

FIGURE 5.5. Zero-field specific heat C of the two-dimensional square Ising model. C diverges logarithmically on both sides of the critical point T_c.

To obtain the "Onsager logarithm" it is not necessary, as shown above, to evaluate the integrals exactly. The integral 5.9 can, however, be evaluated exactly and one finds that

$$U = -J \coth 2v \left[1 + (2 \tanh^2 2v - 1)\frac{2}{\pi} K(k_1)\right], \tag{5.15}$$

where

$$k_1 = \frac{2 \sinh 2v}{\cosh^2 2v}, \tag{5.16}$$

and $K(k_1)$ is the complete elliptic integral of the first kind defined by

$$K(k_1) = \int_0^{\pi/2} (1 - k_1^2 \sin^2 \theta)^{-1/2}\, d\theta. \tag{5.17}$$

The analysis required for Equation 5.15 is straightforward and is left as an exercise for the reader (see Problem 5). The specific heat is obtained from the definition 5.14 and the properties of elliptic integrals (see Problem 6). It is given by

$$C = \frac{2k}{\pi}(v \coth 2v)^2 \Big\{2K(k_1) - 2E(k_1) - 2(1 - \tanh^2 2v) \left[\frac{\pi}{2} + (2\tanh^2 2v - 1)K(k_1)\right]\Big\}, \tag{5.18}$$

where $E(k_1)$ is the complete elliptic integral of the second kind, defined by

$$E(k_1) = \int_0^{\pi/2} (1 - k_1^2 \sin^2 \theta)^{1/2}\, d\theta. \tag{5.19}$$

In the neighborhood of $k_1 = 1-$ (see Problem 7)

$$K(k_1) \sim \log[4(1 - k_1^2)^{-1/2}], \tag{5.20}$$

so from the exact result (Equation 5.18) we obtain a logarithmically divergent specific heat.

As remarked previously, the two-dimensional model in an external field is an unsolved problem. Nevertheless, by various indirect means a number of people have obtained an expression for the spontaneous magnetization, or what is thought to be the spontaneous magnetization. Thermodynamically the spontaneous magnetization is defined by

$$m_0 = \lim_{H \to 0+} \frac{\partial}{\partial(\beta H)}\left(-\frac{\psi}{kT}\right) \tag{5.21}$$

with the limit $H \to 0+$ taken after the thermodynamic limit. An exact evaluation of m_0, therefore, really requires the solution of the $H \neq 0$ problem. Indirect methods based on alternative definitions of spontaneous magnetization, which we shall not go into here, suggest that

$$m_0 = \begin{cases} [1 - (\sinh 2\nu)^{-4}]^{1/8} & T < T_c \\ 0 & T \geq T_c, \end{cases} \tag{5.22}$$

but we stress that nobody has actually proved that this expression agrees with the definition (Equation 5.21).

The expression 5.22 was first derived by Onsager in the middle 1940s, but in true Onsager fashion he has not to this day published his derivation. He tantalized numerous people by writing down Equation 5.22 in various places [Onsager (1949)], and it was not until 1952 that C. N. Yang [Yang (1952a)] gave the first published derivation. Yang's derivation and others given recently [e.g., Montroll et al. (1963)] are extremely complicated. In view of the simplicity of the final result, this fact is both surprising and frustrating.

The two-dimensional problem has been reformulated and solved by many people in a variety of ways, in the hope that a new derivation would light the way for a solution to the three-dimensional problem. This hope has unfortunately not materialized as yet, although many interesting results and interrelations between mathematics and other branches of physics have emerged. Most notable among these is the relation between the Ising problem and combinatorial mathematics. Some new results in combinatorics have been obtained by this pursuit, but unfortunately little has been added to our knowledge of the Ising model. We shall devote Chapter 6 to the combinatorial approach to the Ising problem.

In conclusion we remark that all methods to date can be applied with only mild variations to other two-dimensional lattices (e.g., triangular and hexagonal) and to lattices with different coupling constants in different directions. Although the critical points are different in each case, the critical behavior is the same—a logarithmically divergent specific heat and a $\frac{1}{8}$ power law for the spontaneous magnetization. We shall have more to say about this lattice invariance of critical behavior in Chapter 6.

5-6 Correlation Functions, Eigenvalue Degeneracy, and Long-range Order

The pair-correlation function $\langle \mu_k \mu_l \rangle$ is defined in general by

$$\langle \mu_k \mu_l \rangle = Z_N^{-1} \sum_{\{\mu\}} \mu_k \mu_l \exp(-\beta E\{\mu\}), \tag{6.1}$$

where $E\{\mu\}$, Equation 1.1, is the interaction energy, i.e.,

$$E\{\mu\} = -J \sum_{P,Q}^{*} \mu_P \mu_Q - H \sum_P \mu_P . \tag{6.2}$$

Similarly, one defines three spin-correlation functions $\langle \mu_k \mu_l \mu_m \rangle$, etc.

The one spin-correlation function $\langle \mu_k \rangle$ is essentially the magnetization $(= N^{-1} \sum_k \langle \mu_k \rangle = \langle \mu_k \rangle$ for periodic lattices), which vanishes for finite N when $H = 0$ because of the symmetry of the interaction energy under the transformation $\mu_P \to -\mu_P$, all lattice points P.

In zero field the pair-correlation function is of particular interest since it in a sense measures the "degree of order" of the lattice. Thus if r_{kl} denotes the distance between lattice points k and l and

$$\rho = \lim_{r_{kl} \to \infty} \lim_{N \to \infty} \langle \mu_k \mu_l \rangle \tag{6.3}$$

exists and is nonzero, we say that there is long-range order, which is to say that spins μ_k and μ_l are not independent of one another when separated by an infinite distance.

It seems reasonable that if long-range order exists, there will be a phase transition (i.e., a nonanalytic point of the free energy). This, in fact, must be the case if there is long-range order at (sufficiently) low temperatures and zero long-range order at (sufficiently) high temperatures, since an analytic function cannot be nonzero in one region and identically zero in another region. On the other hand, if there is a nonanalytic point of the free energy, it is conceivable that there will be no long-range order at any temperature. This is indeed the case for an antiferromagnet, where $\langle \mu_k \mu_l \rangle$ oscillates in sign, but for a ferromagnet the question is still open. There is a suggestion that one can have a ferromagnetic transition (divergent susceptibility as $T \to T_c+$ for the two-dimensional Heisenberg model) with zero long-range order at all temperatures. There is also a suggestion that one can have a nonzero spontaneous magnetization without long-range order, but at the moment there is no proof of either statement. Since most of the recent work on the existence and nonexistence of phase transitions in lattice models is based on long-range-order considerations, it would be extremely valuable to settle these questions. The interested reader is referred to the article by Griffiths (1966), where the problems are stated most clearly.

For the one-dimensional model the evaluation of $\langle \mu_k \mu_l \rangle$ is straightforward, particularly for the open chain (see Problem 8). We shall consider here the slightly more complicated closed-chain problem in order to facilitate the discussion of the two-dimensional problem.

In one dimension with periodic boundary conditions (i.e., a closed chain of N spins with $\mu_{N+1} = \mu_1$)

$$\langle \mu_k \mu_l \rangle = Z_N^{-1} \sum_{\{\mu\}} \mu_k \mu_l \exp\left(\nu \sum_{j=1}^{N} \mu_j \mu_{j+1} + B \sum_{j=1}^{N} \mu_j \right), \tag{6.4}$$

where $\nu = J/kT$ and $B = H/kT$. Defining as before (Equation 3.8) the transfer matrix $\mathbf{L}$ with components

$$L(\mu, \mu') = \exp\left[\nu\mu\mu' + \frac{B}{2}(\mu + \mu') \right], \tag{6.5}$$

Equation 6.4 can be written, assuming $l > k$, as

$$\langle \mu_k \mu_l \rangle = Z_N^{-1} \sum_{\{\mu\}} L(\mu_1, \mu_2) \cdots L(\mu_{k-1}, \mu_k)\mu_k L(\mu_k, \mu_{k+1}) \cdots L(\mu_{l-1}, \mu_l)\mu_l L(\mu_l, \mu_{l+1}) \cdots L(\mu_N, \mu_1). \tag{6.6}$$

Summing over all $\mu_j = \pm 1$ in Equation 6.6 except μ_k and μ_l gives

$$\langle \mu_k \mu_l \rangle = Z_N^{-1} \sum_{\substack{\mu_k = \pm 1 \\ \mu_l = \pm 1}} \mu_k L^{N-l+k}(\mu_k, \mu_l)\mu_l L^{l-k}(\mu_l, \mu_k), \tag{6.7}$$

where $L^s(\mu, \mu')$ denotes the (μ, μ') component of the matrix $\mathbf{L}$ raised to the sth power. Now

$$L^s(\mu, \mu') = \sum_{j=1}^{2} \lambda_j^s \phi_j(\mu)\phi_j(\mu'), \tag{6.8}$$

where λ_j and ϕ_j are, respectively, the eigenvalues and corresponding eigenvectors of the matrix $\mathbf{L}$. For example, when $H = 0$ (see Equations 3.10 and 3.16),

$$\lambda_1 = 2\cosh \nu, \qquad \lambda_2 = 2 \sinh \nu, \tag{6.9}$$

and

$$\phi_1 = \begin{pmatrix} \phi_1(+1) \\ \phi_1(-1) \end{pmatrix} = 2^{-1/2}\begin{pmatrix} 1 \\ 1 \end{pmatrix}, \qquad \phi_2 = \begin{pmatrix} \phi_2(+1) \\ \phi_2(-1) \end{pmatrix} = 2^{-1/2}\begin{pmatrix} 1 \\ -1 \end{pmatrix}. \tag{6.10}$$

In general, from Equations 6.7 and 6.8 and the fact that (Equation 3.12)

$$Z_N = \lambda_1^N + \lambda_2^N, \tag{6.11}$$

we have

$$\begin{aligned}\langle \mu_k \mu_l \rangle &= (\lambda_1^N + \lambda_2^N)^{-1} \sum_{\substack{\mu_k = \pm 1 \\ \mu_l = \pm 1}} \sum_{i,j=1}^{2} \lambda_i^{N-l+k} \mu_k \phi_i(\mu_k) \phi_i(\mu_l) \lambda_j^{l-k} \mu_l \phi_j(\mu_k) \phi_j(\mu_l) \\ &= [1 + (\lambda_2/\lambda_1)^N]^{-1} \sum_{i,j=1}^{2} \left(\frac{\lambda_i}{\lambda_1}\right)^{N-l+k} \left(\frac{\lambda_j}{\lambda_1}\right)^{l-k} (\phi_i, \mu\phi_j)^2,\end{aligned} \tag{6.12}$$

where

$$(\phi_i, \mu\phi_j) = \sum_{\mu = \pm 1} \mu \phi_i(\mu) \phi_j(\mu). \tag{6.13}$$

Now for fixed k and l, since $\lambda_2 < \lambda_1$,

$$\lim_{N \to \infty} \left(\frac{\lambda_i}{\lambda_1}\right)^{N-l+k} = \delta_{i,1}. \tag{6.14}$$

Also, since $\lim_{N \to \infty} (\lambda_2/\lambda_1)^N = 0$,

$$\rho_{kl} = \lim_{N \to \infty} \langle \mu_k \mu_l \rangle = \sum_{j=1}^{2} \left(\frac{\lambda_j}{\lambda_1}\right)^{l-k} (\phi_1, \mu\phi_j)^2. \tag{6.15}$$

In zero field, from Equation 6.10,

$$(\phi_1, \mu\phi_j) = \delta_{j,2}. \tag{6.16}$$

Hence, from Equations 6.9 and 6.15,

$$\begin{aligned}\rho_{kl} &= \left(\frac{\lambda_2}{\lambda_1}\right)^{l-k} \\ &= (\tanh \nu)^{l-k} \qquad (l > k).\end{aligned} \tag{6.17}$$

For finite N and $H = 0$ it is easily verified from Equations 6.9, 6.10, and 6.12 that $(l > k)$

$$\langle \mu_k \mu_l \rangle = [1 + (\tanh \nu)^N]^{-1} [(\tanh \nu)^{l-k} + (\tanh \nu)^{N-l+k}]. \tag{6.18}$$

As expected, from Equation 6.17,

$$\lim_{|k-l| \to \infty} \rho_{kl} = \lim_{|k-l| \to \infty} (\tanh \nu)^{|k-l|} = 0 \tag{6.19}$$

for all $\nu > 0$; i.e., there is no long-range order in one dimension for any finite temperature. (At zero temperature corresponding to $\nu \to \infty$ there is, of course, long-range order; in fact, there is complete order.)

In two dimensions the above derivation can be repeated essentially word for word with only slight changes in notation. Thus consider the pair-correlation function $\langle \mu_{k,l}\mu_{k,l+r}\rangle$ of two spins in the kth row and the lth and $(l+r)$th columns, respectively. Since the 2^m by 2^m transfer matrix $L(\sigma, \sigma')$ (Equation 4.7) transfers from one column with configuration $\sigma = (\mu_1, \ldots, \mu_m)$ to the neighboring column with configuration $\sigma' = (\mu_1', \ldots, \mu_m')$ we can write, by analogy with Equations 6.6 and 6.7,

$$\begin{aligned}\langle \mu_{k,l}\mu_{k,l+r}\rangle &= Z_{n,m}^{-1} \sum_{\{\mu\}} \mu_{k,l}\mu_{k,l+r} \exp(-\beta E\{\mu\}) \\ &= Z_{n,m}^{-1} \sum_{\sigma,\sigma'} \mu_k L^r(\sigma, \sigma')\mu_k' L^{n-r}(\sigma', \sigma),\end{aligned} \tag{6.20}$$

where σ and σ' denote, respectively, the configurations of the lth and $(l+r)$th colums. Using (see Equation 6.8)

$$L^s(\sigma, \sigma') = \sum_{j=1}^{2^m} \lambda_j^s \phi_j(\sigma)\phi_j(\sigma'), \tag{6.21}$$

where λ_j and ϕ_j are, respectively, the eigenvalues and corresponding eigenvectors of the matrix $\mathbf{L}$, enables us to write

$$\langle \mu_{k,l}\mu_{k,l+r}\rangle = Z_{n,m}^{-1} \sum_{i,j=1}^{2^m} \lambda_i^{n-r}\lambda_j^r(\phi_i, \mu_k\phi_j)^2, \tag{6.22}$$

where it is to be noted from periodicity that the scalar product does not depend on k. Now since (Equation 4.10)

$$Z_{n,m} = \sum_{j=1}^{2^m} \lambda_j^n \tag{6.23}$$

we have in the limit $n \to \infty$ with r and m fixed,

$$\begin{aligned}C_m(r) &= \lim_{n\to\infty} \langle \mu_{k,l}\mu_{k,l+r}\rangle \\ &= \sum_{j=1}^{2^m} \left(\frac{\lambda_j}{\lambda_1}\right)^r (\phi_1, \mu_k\phi_j)^2.\end{aligned} \tag{6.24}$$

Since the principal eigenvector ϕ_1 in zero field is symmetric [i.e., $\phi_1(\sigma) = \phi_1(-\sigma)$], the $j = 1$ term in Equation 6.24 vanishes, so the question of the existence of long-range order is now a question of the degeneracy of the maximum eigenvalue λ_1 of $\mathbf{L}$. For finite m we can appeal to a theorem of

Frobenius [see Gantmacher (B1964)], which states that the maximum eigenvalue of a finite matrix whose components are all positive (see Equation 4.7) is strictly nondegenerate. It then follows that there is no long-range order for an infinite ($n \to \infty$) by finite (m) lattice. That is, from Equation 6.24, since the $j = 1$ term vanishes and $(\lambda_j/\lambda_1) < 1$ for all $j \geq 2$,

$$\lim_{\substack{r\to\infty \\ (m \text{ fixed})}} C_m(r) = 0. \tag{6.25}$$

In the limit $m \to \infty$, however, as shown in Appendix D, λ_1 becomes asymptotically degenerate for $T < T_c$ (i.e., for temperatures below the critical temperature). Precisely, we have

$$\begin{aligned} \frac{\lambda_2}{\lambda_1} &< 1 && \text{for all } m \text{ and } T > T_c \\ &= 1 - 0(e^{-cm}) && \text{as } m \to \infty \text{ and } T < T_c. \end{aligned} \tag{6.26}$$

It follows from Equation 6.24 that

$$\lim_{r\to\infty} \lim_{m\to\infty} C_m(r) = \begin{cases} \lim\limits_{m\to\infty} (\phi_1, \mu_k \phi_2)^2 > 0 & \text{for } T < T_c \\ 0 & \text{for } T > T_c; \end{cases} \tag{6.27}$$

i.e., there is long-range order below T_c and zero long-range order above T_c, as one might have expected.

It is clear that, in general, whenever one has a formula such as Equation 6.24 that long-range order exists if and only if the maximum eigenvalue λ_1 is asymptotically degenerate [i.e., $\lim_{m\to\infty} (\lambda_2/\lambda_1) = 1$ and $(\phi_1, \mu\phi_2) \neq 0$ for all m]. Thus from Equation 6.24

$$\begin{aligned} C_m(r) &\leq \left(\frac{\lambda_2}{\lambda_1}\right)^r \sum_{j=1}^{2m} (\phi_1, \mu\phi_j)^2 \\ &= \left(\frac{\lambda_2}{\lambda_1}\right)^r (\phi_1, \mu^2\phi_1)^2 \\ &= \left(\frac{\lambda_2}{\lambda_1}\right)^r, \end{aligned} \tag{6.28}$$

where in the second step we have made use of Parseval's theorem. It follows from Equation 6.28 that if λ_1 is not asymptotically degenerate, there is zero long-range order. Similarly, from Equation 6.24),

$$C_m(r) \geq \left(\frac{\lambda_2}{\lambda_1}\right)^r (\phi_1, \mu\phi_2)^2, \tag{6.29}$$

it follows that there is long-range order if λ_1 is asymptotically degenerate and $\lim_{m\to\infty}(\phi_1, \mu\phi_2) \neq 0$, which proves the assertion.

It has been suggested by Kac (1968) that the phenomenon of eigenvalue degeneracy may well be a general mathematical mechanism for a phase transition to long-range order. The idea is to construct from the Hamiltonian of the system under consideration an operator whose largest eigenvalue gives the (limiting) free energy of the systems, such that a phase transition occurs if and only if the largest eigenvalue is asymptotically degenerate. This program has been carried through for a number of model systems [e.g., Kac (1968) and Thompson (1968a)], so it may well turn out that eigenvalue degeneracy provides a general mathematical mechanism for phase transitions.

To conclude this section we present a physical interpretation of the principal eigenvector ϕ_1.

Consider a particular column of the lattice and denote by $P(\sigma)$ the probability that the column is in a particular configuration σ. By definition

$$P(\sigma) = Z_{n,m}^{-1} \sum_{\{\mu\}}{}' \exp(-\beta E\{\mu\}), \tag{6.30}$$

where the primed sum is over all configurations of the lattice with the column configuration σ held fixed. In terms of the transfer matrix (see Equation 4.6) we can write

$$P(\sigma) = \frac{L^n(\sigma, \sigma)}{\mathrm{Tr}(\mathbf{L}^n)}, \tag{6.31}$$

and in view of Equations 4.10 and 6.21 (with $\sigma' = \sigma$) we have

$$\lim_{n\to\infty} P(\sigma) = \phi_1^2(\sigma). \tag{6.32}$$

$\phi_1^2(\sigma)$, therefore, provides us with another measure of the degree of order of the lattice. Physically one would expect $\phi_1^2(\sigma)$ to be concentrated around "ordered configurations" at low temperatures ($T < T_c$) and equally spread among all configurations at high temperatures ($T > T_c$).

One final point to note is that we have considered above only pair-correlation functions for two spins in the same row. We could equally well have considered the pair-correlation function for two spins in the same column. The final result for this situation (see Problem 9) is

$$\lim_{n\to\infty} \langle \mu_{k,l}\, \mu_{k+r,l} \rangle = \sum_{\sigma} \phi_1(\sigma) \mu_k \mu_{k+r} \phi_1(\sigma), \tag{6.33}$$

where $\sigma = (\mu_1, \ldots, \mu_m)$ denotes the configuration of the lth column. Eigenvalue degeneracy now does not seem to play any role in the discussion of long-range column order. It does, however, appear in a devious way through the ordered–disordered form of $\phi_1^2(\sigma)$ described above. Note also that by symmetry, Equations 6.33 and 6.24 are identical in the limit $m, n \to \infty$. This rather peculiar identity seems to be extremely difficult to prove directly.

PROBLEMS

1. Consider the lattice shown in Figure 5.6 with spins $\mu_P = \pm 1$ on the (six) vertices. What is the minimum (i.e., antiferromagnetic ground state) of the expression $\sum_{P,Q}^{*} \mu_P \mu_Q$, where the sum is over nearest-neighbor lattice points P and Q, and for what configurations is the minimum achieved?

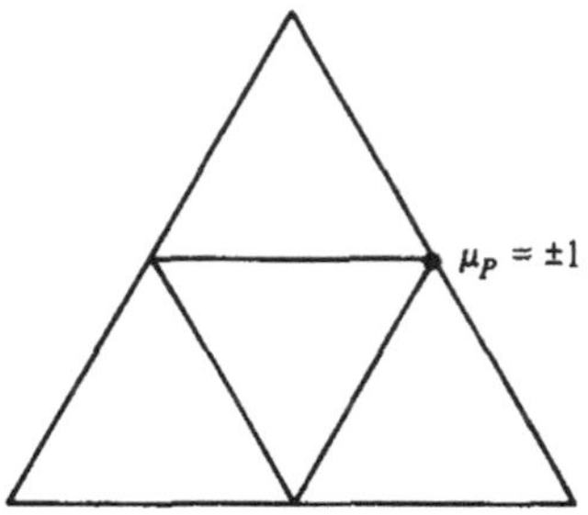

FIGURE 5.6. Six-spin antiferromagnetic Ising lattice.

2. Derive the relations between susceptibility and compressibility, Equation 2.20, and between specific heats, Equation 2.21 for the Ising magnet and the lattice gas.

3. Compute the grand-canonical partition function $Z_G(z, V, T)$ for the one-dimensional lattice gas with interaction energy

$$E\{t\} = -J \sum_{i=1}^{V} t_i t_{i+1},$$

where $t_i = 1$ or 0, $t_{V+1} = t_1$, and $\sum_{i=1}^{V} t_i = N$, and

(a) show that for $J > 0$ the zeros of $Z_G(z, V, T)$ lie on the unit circle in the complex z plane.

(b) Compute the pressure as a function of z.

4. Consider a 2 by N lattice with the Nth column coupled to the first and with interaction energy

$$E\{\mu, \mu'\} = -J\sum_{i=1}^{N} \mu_i \mu_i' - J\sum_{i=1}^{N} (\mu_i \mu_{i+1} + \mu_i' \mu_{i+1}')$$

($\mu_{N+1} = \mu_1$ and $\mu_{N+1}' = \mu_1'$) for the configuration shown in Figure 5.7. By transferring from one column to the next, show that the partition function $Z_{2,N}$ can be written as

$$Z_{2,N} = \sum_{\{\mu, \mu'\}} \exp(-\beta E\{\mu, \mu'\}) = \mathrm{Tr}(\mathbf{A}^N),$$

where

$$\mathbf{A} = \begin{pmatrix} e^{3\nu} & 1 & 1 & e^{-\nu} \\ 1 & e^{\nu} & e^{-3\nu} & 1 \\ 1 & e^{-3\nu} & e^{\nu} & 1 \\ e^{-\nu} & 1 & 1 & e^{3\nu} \end{pmatrix}, \qquad \nu = \beta J.$$

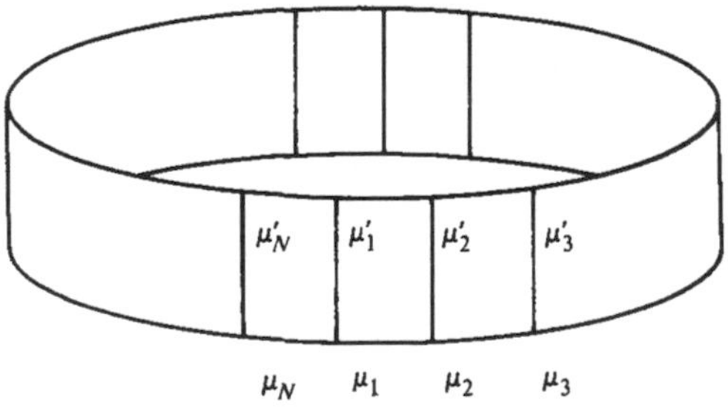

FIGURE 5.7. A 2 by N Ising model.

5. By performing one of the θ_i integrations in Equation 5.9 for the internal energy U of the two-dimensional Ising model, derive Equation 5.15 for U in terms of the complete elliptic integral of the first kind,

$$K(k) = \int_0^{\pi/2} (1 - k^2 \sin^2 \theta)^{-1/2}\, d\theta.$$

6. Derive the result,

$$k\frac{d}{dk}K(k) = (1 - k^2)^{-1}E(k) - K(k),$$

where

$$E(k) = \int_0^{\pi/2} (1 - k^2 \sin^2 \theta)^{1/2}\, d\theta$$

is the complete elliptic integral of the second kind.

7. By writing

$$K(k) = \int_0^{\pi/2} (1 - k \sin \theta)(1 - k^2 \sin^2 \theta)^{-1/2}\, d\theta + \int_0^{\pi/2} k \sin \theta (1 - k^2 \sin^2 \theta)^{-1/2}\, d\theta,$$

show that

$$K(k) \sim \log[4(1 - k^2)^{-1/2}] \qquad \text{as } k \to 1-.$$

8. By summing successively over the Nth, $(N-1)$th, etc., spins ($\mu_N = \pm 1$, $\mu_{N-1} = \pm 1$, etc.), show that

$$Z(J_1, J_2, \ldots, J_{N-1}) = \sum_{\{\mu\}} \exp\left(\sum_{i=1}^{N-1} J_i \mu_i \mu_{i+1}\right) = 2 \prod_{i=1}^{N-1} (2 \cosh J_i).$$

By differentiating with respect to $J_k, J_{k+1}, \ldots, J_{k+r}$ and setting $J_i = J$ all $i = 1, 2, \ldots, N-1$, show that the two-spin correlation function for the open chain is given by

$$\langle \mu_k \mu_{k+r} \rangle = [Z(J, \ldots, J)]^{-1} \sum_{\{\mu\}} \mu_k \mu_{k+r} \exp\left(J \sum_{i=1}^{N-1} \mu_i \mu_{i+1}\right) = (\tanh J)^r \qquad \text{for all } N > 1.$$

9. Derive Equation 6.33 for the pair-correlation function of two spins in a column.

CHAPTER 6

The Ising Model: Combinatorial Approach

6-1 Formulation

The basic idea behind the combinational approach is extremely simple and was noted many years ago by Van der Waerden (1941). The problem is to evaluate the partition function Z_N (Equation 1.7 of Chapter 5), which can be written in the form

$$Z_N = \sum_{\{\mu\}} \prod_{P,Q}{}^{*} \exp(\nu \mu_P \mu_Q), \tag{1.1}$$

where the starred product is over nearest-neighbor lattice points P and Q, $\nu = J/kT$, and the sum is over all configurations of the lattice ($\mu_P = \pm 1$). Expanding the exponential in Equation 1.1 and noting, since $\mu_P^2 = +1$, that

$$(\mu_P \mu_Q)^n = \begin{cases} 1 & \text{if } n \text{ is even} \\ \mu_P \mu_Q & \text{if } n \text{ is odd,} \end{cases} \tag{1.2}$$

we have

$$\begin{aligned} \exp(\nu \mu_P \mu_Q) &= \cosh \nu + \mu_P \mu_Q \sinh \nu \\ &= \cosh \nu (1 + \omega \mu_P \mu_Q), \end{aligned} \tag{1.3}$$

where

$$\omega = \tanh \nu. \tag{1.4}$$

We can then write the partition function, Equation 1.1, in the form

$$Z_N = (\cosh v)^{Nq/2} \sum_{\{\mu\}} \prod_{P,Q}{}^{*} (1 + \omega\mu_P \mu_Q), \tag{1.5}$$

where q is the lattice coordination number and N is the number of lattice sites (so that $\sum^{*}_{P,Q} = Nq/2$ is the total number of nearest neighbor bonds). We now expand the product in Equation 1.5,

$$\prod_{P,Q}{}^{*} (1 + \omega\mu_P \mu_Q) = 1 + \omega \sum_{P,Q}{}^{*} \mu_P \mu_Q + \omega^2 \sum_{\substack{P_1,Q_1 \\ (P_1,Q_1)\neq(P_2,Q_2)}}{}^{*} \sum_{P_2,Q_2}{}^{*} \mu_{P_1}\mu_{Q_1}\mu_{P_2}\mu_{Q_2} + \cdots \tag{1.6}$$

and represent each product $\mu_P \mu_Q$ corresponding to a nearest-neighbor pair of lattice points P and Q by a bond on the lattice connecting P and Q. For example, each term in the coefficient of ω has the representation shown in Figure 6.1 and each term in the coefficient of ω^2 has either the representation (a) or (b) shown in Figure 6.2. In case (a),

$$\mu_{P_1}\mu_{Q_1}\mu_{P_2}\mu_{Q_2} = \mu_{P_1}(\mu_{Q_1})^2\mu_{Q_2} = \mu_{P_1}\mu_{Q_2} \tag{1.7}$$

and in general any point that appears an even number of times in a product of μ's has its μ^n replaced by 1. Similarly, any point that appears an odd number of times has a μ remaining. Since

$$\sum_{\mu=\pm 1} \mu = 0 \quad \text{and} \quad \sum_{\mu=\pm 1} \mu^2 = 2, \tag{1.8}$$

it then follows from Equations 1.5 and 1.6 that

$$Z_N = 2^N(\cosh v)^{Nq/2} \sum_{r=0}^{\infty} n(r)\omega^r, \tag{1.9}$$

where $n(0) = 1$ and $n(r)$ is the number of graphs that can be constructed from r bonds on the lattice with the restrictions that (1) no bond can occur more

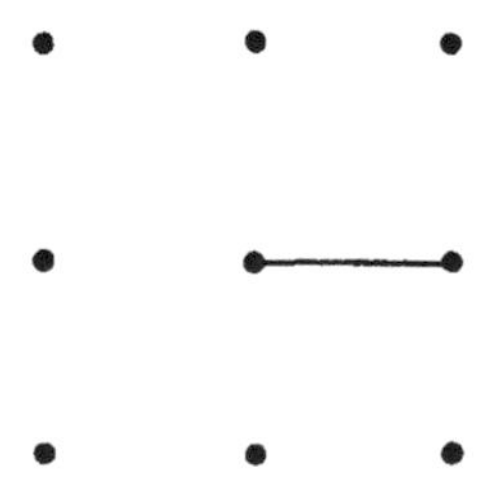

FIGURE 6.1. Possible one-bond graphs for the two-dimensional Ising model.

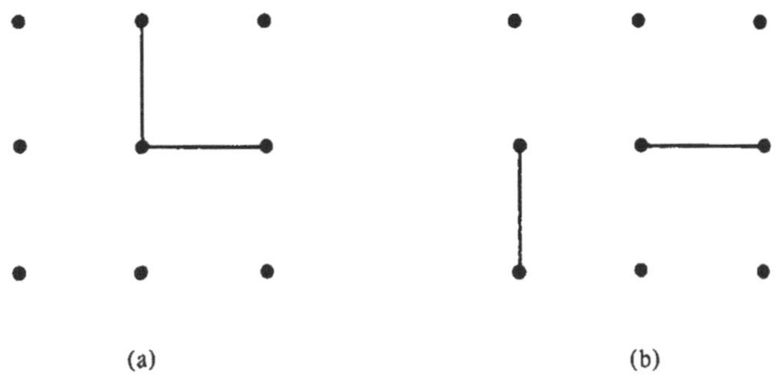

FIGURE 6.2. Possible two-bond graphs for the two-dimensional Ising model.

than once in a given graph, and (2) only an even number of bonds can meet at a lattice point (e.g., 0, 2, or 4 for the square lattice). The problem is therefore reduced to counting graphs on a lattice. For simplicity we shall refer to graphs satisfying conditions (1) and (2) as closed graphs.

This approach is clearly ideally suited to generating series expansions since one has merely to count graphs. The number of graphs $n(r)$, of course, becomes enormous as r increased, particular for three-dimensional lattices. Nevertheless, a relatively large number of terms have been calculated (15 or so in three dimensions). These expansions are our only real source of exact information in three dimensions. We shall have more to say about this in Section 6-5.

To illustrate the counting procedure let us consider the two-dimensional square lattice wrapped on a torus. Clearly, there can be no graphs satisfying conditions (1) and (2) for $r = 1$, 2, or 3. For $r = 4$ the unit square shown in Figure 6.3 is the only possible graph, and since it can be in any one of N places,

$$n(4) = N; \tag{1.10}$$

in other words, there is one square per site. When $r = 5$ there are no graphs satisfying the required conditions, and in fact the reader can easily convince himself that

$$n(r) = 0 \qquad \text{for } r \text{ odd.} \tag{1.11}$$

When $r = 6$ there are two possible graphs shown in Figure 6.4 and since there are N of each,

$$n(6) = 2N; \tag{1.12}$$

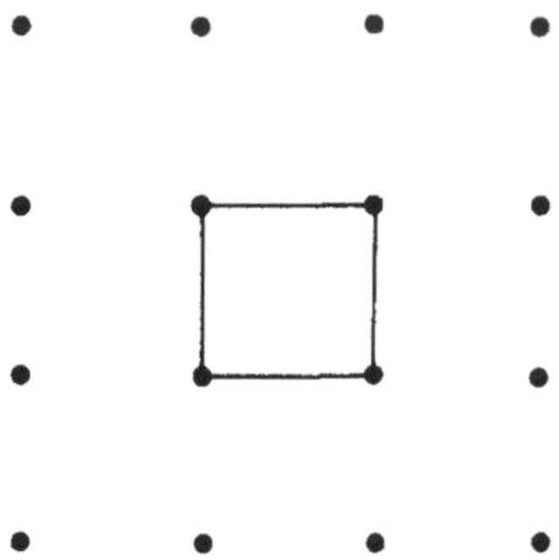

FIGURE 6.3. Only contributing four-bond graph.

or

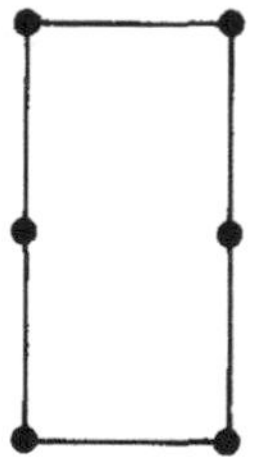

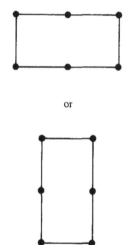

FIGURE 6.4. Contributing six-bond graphs.

in other words, there are two hexagons per site. When $r = 8$ we can have the connected graphs shown in Figure 6.5, giving a total or $7N$ graphs, or a disconnected graph consisting of two disjoint squares. With one square fixed there are $N - 5$ possible positions for the other square, giving a total contribution of $\frac{1}{2}N(N-5)$. It follows that

$$\begin{aligned} n(8) &= 7N + \tfrac{1}{2}N(N-5) \\ &= \tfrac{1}{2}N^2 + \tfrac{9}{2}N. \end{aligned} \tag{1.13}$$

For larger r the problem rapidly becomes complicated, so we shall stop at this point.

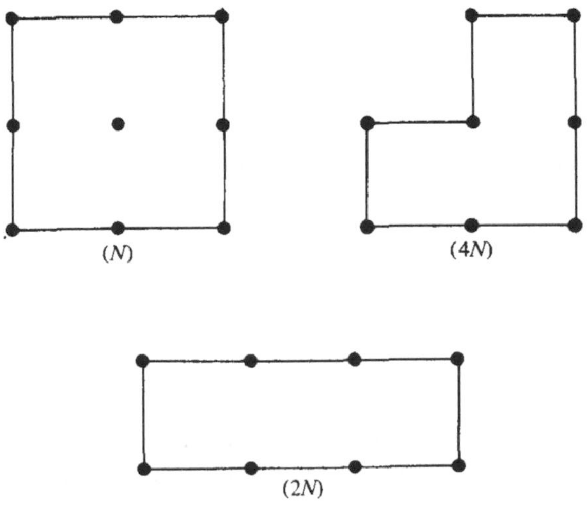

FIGURE 6.5. Contributing connected eight-bond graphs.

From Equation 1.9 to 1.13 we have that $(q = 4)$

$$Z_N = (2\cosh^2 v)^N[1 + N\omega^4 + 2N\omega^6 + (\tfrac{1}{2}N^2 + \tfrac{9}{2}N)\omega^8 + \cdots]. \tag{1.14}$$

For large N, $\log Z_N$ should be proportional to N (since we know from Chapter 5 that $\lim_{N\to\infty} N^{-1} \log Z_N$ exists). To see how this comes about from Equation 1.14 we take logarithms of both sides of Equation 1.14 and formally expand the logarithm of the bracketed term. If we do this and retain only terms up to ω^8 we obtain

$$\begin{aligned}\log Z_N &= N \log 2\cosh^2 v + N\omega^4 + 2N\omega^6 + (\tfrac{1}{2}N^2 + \tfrac{9}{2}N)\omega^8 - \tfrac{1}{2}(N\omega^4)^2 + \cdots \\ &= N(\log 2\cosh^2 v + \omega^4 + 2\omega^6 + \tfrac{9}{2}\omega^8 + \cdots),\end{aligned} \tag{1.15}$$

which has the required form. The important point to note is that the N^2 term in Equation 1.14 cancels when one computes the free energy. The expansion of the logarithm of Equation 1.14 is not legitimate, since the series diverges for sufficiently large N. One obtains the same result, however, by taking

$$Z_N \sim \left(1 + \sum_{n=1}^{\infty} a_n \omega^n\right)^N, \tag{1.16}$$

expanding the right-hand side in a binomial series and equating coefficients with those in Equation 1.14. This process can be made perfectly rigorous.

In general, if r is less than the circumference of the torus (i.e., $r < N^{1/d}$ in d

dimensions), $n(r)$ is a polynomial of degree $m \leq r/4$ (or $\leq r/3$ if triangles can occur), i.e.,

$$n(r) = Na_r^{(1)} + N^2 a_r^{(2)} + \cdots + N^m a_r^{(m)}. \tag{1.17}$$

Taking logarithms as above, the coefficients of $N^2, \ldots, N^m$ vanish and we are left with

$$\begin{aligned} -\frac{\psi}{kT} &= \lim_{N\to\infty} N^{-1} \log Z_N \\ &= \log 2 + \frac{q}{2} \log \cosh \nu + \sum_{r=0}^{\infty} a_r^{(1)} \omega^r. \end{aligned} \tag{1.18}$$

The problem then is to compute $a_r^{(1)}$ in Equation 1.17. More detailed discussions of the counting problem can be found in the review articles by Domb (1960) and Fisher (1965, 1967).

An exact evaluation of Z_N in two dimensions based on the combinational formula 1.9 was first attempted by Kac and Ward (1952). Their idea was to express the generating function for the graph-counting problem (i.e., the partition function 1.9) as a determinant of a matrix. They succeeded in constructing a matrix to give the Onsager result, but they were unable to prove that all graphs were counted correctly (although clearly " most " graphs must have been counted correctly). Sherman (1960) showed that not all graphs were counted correctly by the Kac–Ward matrix, but with the aid of a conjecture of Feynmann, which Sherman himself proved, he was able to make the Kac–Ward argument completely rigorous.

A variant of the Kac–Ward method emerged in the early 1960s when a number of Ising-model enthusiasts became aware of Pfaffians, which were known to mathematicians last century but were subsequently forgotten. This method will be discussed in Section 6-3, where the Ising problem is related to a dimer problem that can be solved by Pfaffians.

6-2 Low-temperature Expansions and Lattice Duality in Two Dimensions

Since $\omega = \tanh(J/kT)$ is small when T is large, the combinational formula 1.9 gives a high-temperature expansion for the partition function. Low-temperature expansions can be developed in a similar manner if we start from the completely ordered state (corresponding to zero temperature), i.e., when all spins are up or all spins are down (for $H = 0$), and successively flip neighboring spins. The first term in the low-temperature expansion for Z_N with $H = 0$ is, then, from Equations 1.6 and 1.7 of Chapter 5,

$$2\exp(\tfrac{1}{2}Nvq), \tag{2.1}$$

the "2" in front coming from the all-up or all-down symmetry of the ground state. The next term is obtained by flipping one spin. The energy of such a configuration is

$$-\tfrac{1}{2}NqJ + 2qJ \tag{2.2}$$

and there are $2N$ such configurations, so the first two terms in the low-temperature expansion for Z_N are given by

$$Z_N = 2\exp(\tfrac{1}{2}Nvq)(1 + N\exp(-2vq) + \cdots). \tag{2.3}$$

In general we can write

$$Z_N = 2\exp(\tfrac{1}{2}Nvq)\sum_{r=0}^{\infty} m(r)\exp(-2vr), \tag{2.4}$$

where $m(0) = 1$ and $m(r)$ is the number of ways r pairs of unlike neighbors can be arranged on the lattice. The generalization of this formula for nonzero magnetic fields is the subject of Problem 2.

We now introduce the notion of a dual lattice. For the square lattice shown in Figure 6.6, the dashed square lattice is the dual lattice (the square lattice is "self-dual"). In general, the dual of a given lattice is obtained by drawing the perpendicular bisector through each bond of the original lattice and connecting the new bonds at points in the center of each unit cell of the original lattice. For example, as shown in Figure 6.7, the dual of the triangular lattice is the hexagonal lattice and vice-versa.

In a given configuration of up spins (denoted by +) and down spins (denoted by −) dashed lines are drawn through each bond connecting an up and a down spin. A typical situation is shown in Figure 6.8 for the square lattice. In general, it is clear that *broken bonds on the original lattice give closed graphs on the dual lattice*, and, in fact, there is a one-to-one correspondence between broken bonds on the original lattice and closed graphs on the dual lattice. In other words, from the definitions of $m(r)$ and $n(r)$,

$$n(r) = m_D(r) \qquad \text{and} \qquad m(r) = n_D(r), \tag{2.5}$$

where $m_D(r)$ denotes the number of ways r pairs of unlike neighbors can be

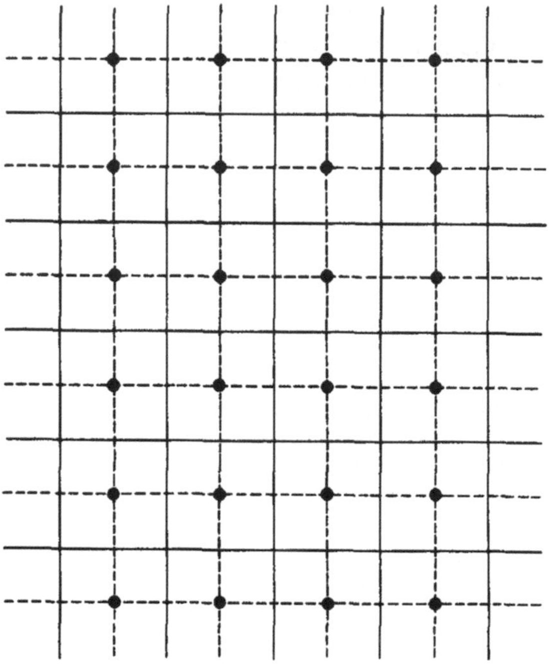

FIGURE 6.6. Square lattice and its dual.

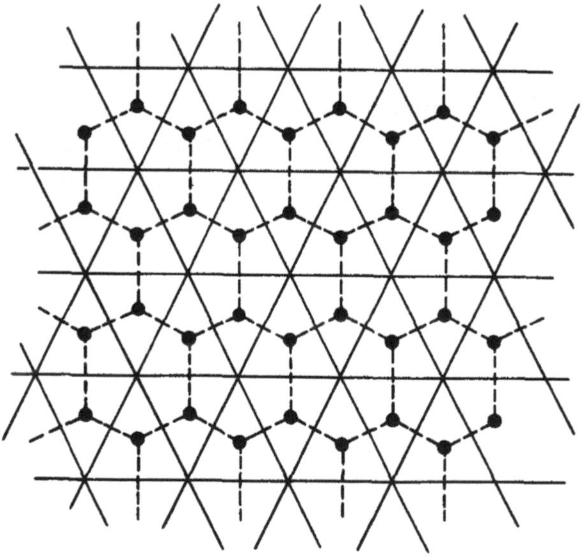

FIGURE 6.7. Triangular lattice and its hexagonal dual lattice.

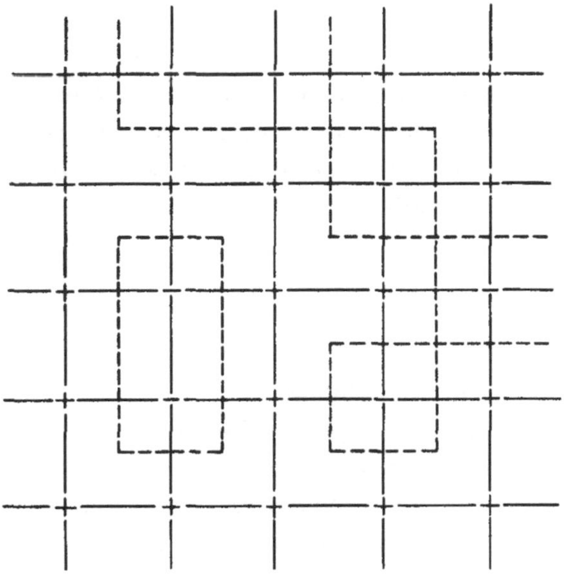

FIGURE 6.8. Closed graphs on the dual lattice correspond to pairs of unlike spins on the original lattice.

arranged on the dual lattice and $n_D(r)$ the number of closed graphs on the dual lattice with r bonds.

If we now define v^* by

$$\tanh v^* = e^{-2v}, \tag{2.6}$$

it is obvious from Equations 2.4 and 1.9 that (exhibiting the temperature dependence)

$$\begin{aligned} Z(v)\exp(-\tfrac{1}{2}Nvq) &= 2\sum_{r=0}^{\infty} m(r)\exp(-2vr) \\ &= 2\sum_{r=0}^{\infty} n_D(r)(\tanh v^*)^r \\ &= 2Z_D(v^*)[2^{N_D}(\cosh v^*)^{N_D q_D/2}]^{-1}, \end{aligned} \tag{2.7}$$

where Z_D denotes the partition function of the dual lattice and N_D and q_D denote the number of lattice points and the coordination number of the dual lattice, respectively. In general there is a bond of the dual lattice for each bond of the original lattice, i.e., $N_D q_D = Nq$, so from Equations 2.6 and 2.7 it follows that

$$Z(v) = Z_D(v^*)2^{1-N_D}(2 \sinh 2v)^{Nq/4}. \tag{2.8}$$

Since the square lattice is self-dual, $Z_D = Z$, so from Equation 2.6, Equation 2.8 relates the square-lattice partition function at a low temperature (large v) to the square-lattice partition at a conjugate high temperature (small v^*) and vice-versa. It follows that singularities must occur in pairs, and if there is one, and only one, singularly it must occur when $v = v^*$, i.e., from Equation 2.6 when

$$\tanh v = e^{-2v} \qquad \text{or} \qquad \sinh 2v = 1, \tag{2.9}$$

which is precisely the equation (Equation 5.12 of Chapter 5) determining the critical point in the Onsager solution.

Unfortunately, the above argument does not work in three dimensions, since there is no three-dimensional analogue of the dual lattice. Equation 2.8 also does not locate the critical points of the triangular and hexagonal lattices. Onsager, however, using a slightly different trick (the "star-triangle transformation") was able to relate the low-temperature expansion for the triangular-lattice partition function to the low-temperature expansion for the hexagonal-lattice partition function. This, combined with the duality relation 2.8, locates the critical points for both lattices, which are given by

$$(\tanh v_c)^{-1} = \begin{cases} 2 + \sqrt{3} & \text{triangular} \\ \sqrt{3} & \text{hexagonal.} \end{cases} \tag{2.10}$$

These ideas have been generalized by Fisher (1959a), who has located the critical points of all sorts of interesting-looking lattices.

Historically, the relation 2.8 was first deduced by Kramers and Wannier (1941) from symmetry properties of the transfer matrix for a (two-dimensional) helical lattice. In this way they deduced the exact critical point (2.9) three years before the appearance of Onsager's solution. The lattice duality argument given above is due to Onsager [see Wannier (1945)].

6-3 Dimer Solution of the Ising Model in Two Dimensions

The dimer problem seems first to have been considered by Fowler and Rushbrooke (1937), who attempted to calculate the entropy of a gas of diatomic molecules absorbed on a crystalline surface.

A simplified version of this problem in two dimensions is to find the number of ways of covering a square lattice with dimers, a dimer being an object that covers bonds connecting nearest-neighbor lattice sites, with the restriction that a given lattice site cannot be occupied by more than one dimer.

To solve this problem it is convenient to consider the generating function defined by

$$\Phi(z_1, z_2) = \sum_{\substack{n_1, n_2 \\ (n_1 + n_2 = N/2)}} g(n_1, n_2) z_1^{n_1} z_2^{n_2}, \tag{3.1}$$

where $g(n_1, n_2)$ is the number of dimer configurations with n_1 horizontal dimers and n_2 vertical dimers, and N is the total number of lattice sites (assumed to be even). The total number of dimer configurations is clearly $\Phi(1, 1)$, so the problem will be solved once we have evaluated Equation 3.1.

The generating function, Equation 3.1, is similar in form to the partition function of the two-dimensional rectangular Ising model. Thus if spins in rows have interaction constant J_1 and spins in columns have interaction constant J_2, a straightforward generalization of the argument leading to Equation 1.9 gives

$$Z_N = (2 \cosh \nu_1 \cosh \nu_2)^N \sum_{r, s=0}^{\infty} n(r, s) \omega_1^r \omega_2^s, \tag{3.2}$$

where $n(r, s)$ is the number of closed graphs with r horizontal bonds and s vertical bonds, and

$$\nu_i = \frac{J_i}{kT}, \quad \omega_i = \tanh \nu_i, \qquad i = 1, 2. \tag{3.3}$$

The trick now is to set up a one-to-one correspondence between closed graphs on the square lattice and dimer configurations on another, appropriate two-dimensional "dimer lattice." The simplest dimer lattice found to date [Fisher (1966)] is the rather decorative one shown in Figure 6.9. Its simplicity comes from the fact that the dimer configurations corresponding to closed graphs are planar; i.e., there are no overlapping dimers (see Figure 6.11). Kasteleyn (1963), who was the first to relate the Ising and dimer problems, constructed a dimer lattice having overlapping dimer configurations which caused considerable difficulties.

We now set up a one-to-one correspondence between every possible closed-graph configuration on the square lattice and dimer configurations on Fisher's dimer lattice (Figure 6.9).

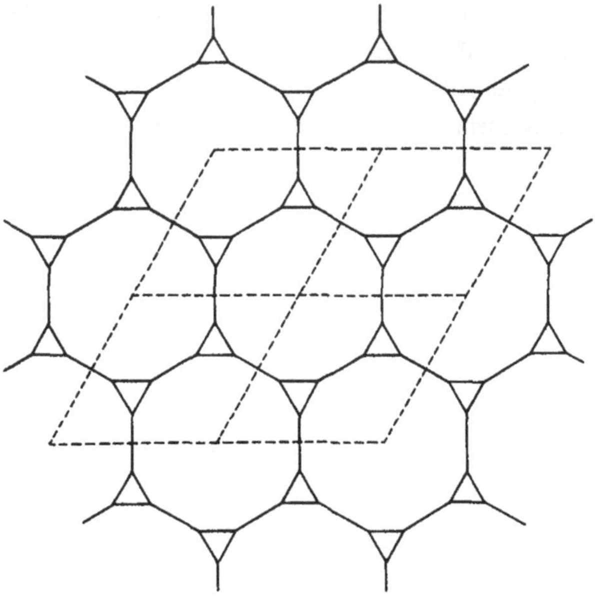

FIGURE 6.9. Fisher's dimer lattice.

The first thing to note is that the square and dimer lattices can be built out of the unit cells shown in Figure 6.10. The unit cells in both cases have four "terminals" and the bonds of the original lattices are now represented by dashed lines in the figure.

If the dimer lattice has the same number of unit cells as the square lattice it is obvious that a one-to-one correspondence can be set up between the unit cells of the two lattices. Moreover, we can build up any closed graph on the

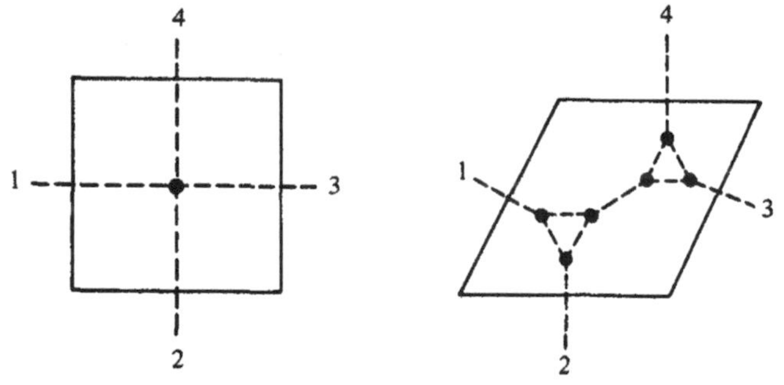

FIGURE 6.10. Unit cells for the square lattice and Fisher's dimer lattice.

square lattice from the eight distinct unit-cell types shown in Figure 6.11, the solid lines in the unit cell now corresponding to half a graph bond on the lattice. We then set up a one-to-one correspondence between bonded square-lattice unit cells and dimer configurations on the dimer-lattice unit cells as shown in Figure 6.11. In view of the one-to-one correspondence between cells and cell configurations, we can then set up a one-to-one correspondence between closed graphs on the square-lattice and dimer configurations on Fisher's dimer lattice. It follows from Equation 3.2 that if we assign a weight ω_1 to dimers passing through terminals 1 or 3 (see Figure 6.10), a weight ω_2 to dimers passing through terminals 2 or 4, and a weight 1 to the interior dimers,

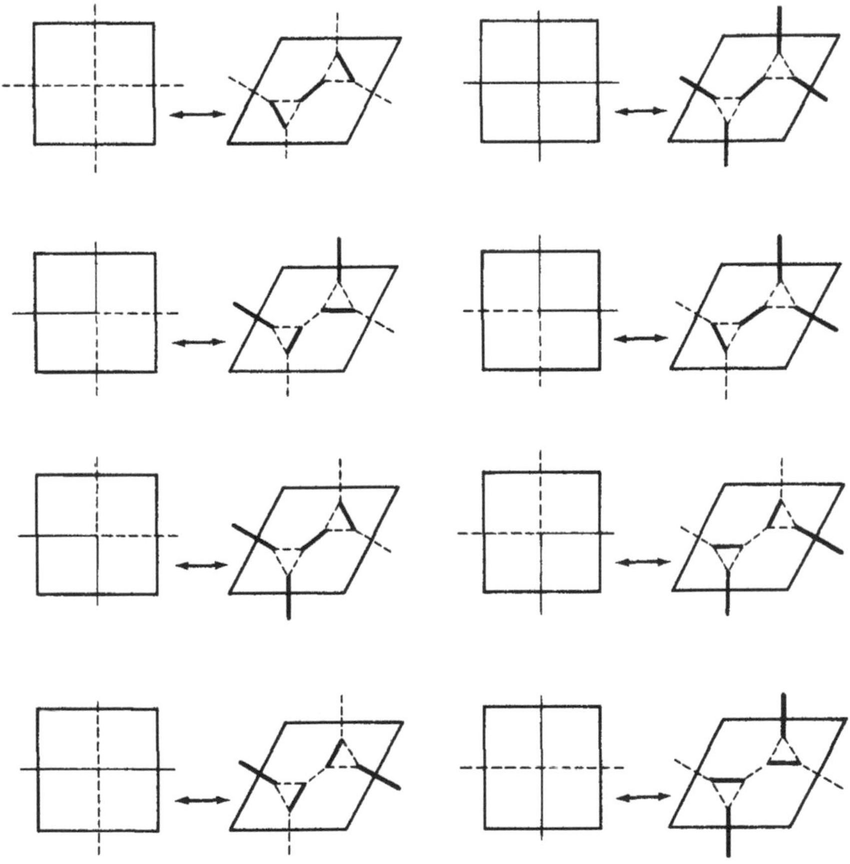

FIGURE 6.11. One-to-one correspondence between bond configurations on the square lattice and dimer configurations on Fisher's lattice.

the partition-function equation (3.2) can be related to the dimer-generating function $\Phi^F(\omega_1, \omega_2)$ for Fisher's lattice through

$$Z_N = (2 \cosh v_1 \cosh v_2)^N \Phi^F(\omega_1, \omega_2). \tag{3.4}$$

The relation between the dimer and Ising problems was first noted by Kasteleyn (1963). Previously, Kasteleyn (1961) and Temperley and Fisher (1961), independently, solved the dimer problem with the aid of Pfaffians. Pfaffians are interesting objects (see Appendix E) which were known to mathematicians last century [Thomson and Tait (B1879)], but since they appeared to be mathematically uninteresting they were rapidly forgotten. Pfaffians were revived in the 1950s by Caianello, who was busy evaluating vacuum expectation values in quantum field theory. It turned out that Pfaffians were ideally suited for the job. Hurst and Green (1960) quickly saw a connection and reevaluated Onsager's solution via vacuum expectation values and Pfaffians. A full account of their work can be found in Green and Hurst (B1964). Almost simultaneously, Kasteleyn, and Fisher and Temperley, solved the dimer problem with the aid of Pfaffians.

Solutions of the dimer problems on a square lattice, and Fisher's lattice, which gives the Ising-model partition function Equation 3.4, are given in Appendix E.

The final result for the number of dimer configurations D on a square lattice is

$$D \sim \exp\left(\frac{2NG}{\pi}\right) \qquad \text{as } N \to \infty, \tag{3.5}$$

where N is the number of lattice sites and G is Catalan's constant:

$$\begin{aligned} G &= 1 - 3^{-2} + 5^{-2} - 7^{-2} + \cdots \\ &= 0.915965594\ldots. \end{aligned} \tag{3.6}$$

The final result for the partition function Equation 3.4 (see Equation 5.8 of Chapter 5) is

$$\lim_{N\to\infty} N^{-1} \log Z_N = \log 2 + (2\pi)^{-2} \iint_0^{2\pi} \log(\cosh 2v_1 \cosh 2v_2 - \sinh 2v_1 \cos \theta_1 - \sinh 2v_2 \cos \theta_2)\, d\theta_1\, d\theta_2. \tag{3.7}$$

6-4 Correlation Functions and Susceptibility

The pair-correlation function $\langle \mu_k \mu_l \rangle$ in zero field is defined by (Equation 6.1 of Chapter 5)

$$\langle \mu_k \mu_l \rangle = Z_N^{-1} \sum_{\{\mu\}} \mu_k \mu_l \exp\left(v \sum_{P,Q}^{*} \mu_P \mu_Q \right). \tag{4.1}$$

In the algebraic formulation of the problem $\langle \mu_k \mu_l \rangle$ is expressed in terms of the eigenvalues and eigenvectors of the transfer matrix (Equations 6.15 and 6.24 of Chapter 5). In the combinational approach we formulate $\langle \mu_k \mu_l \rangle$ as a graph-counting problem, as follows.

Expanding the exponentials in Equation 4.1, making use of Equation 1.3, we obtain

$$\langle \mu_k \mu_l \rangle = Z_N^{-1} (\cosh v)^{Nq/2} \sum_{\{\mu\}} \mu_k \mu_l \prod_{P,Q}^{*} (1 + \mu_P \mu_Q \tanh v). \tag{4.2}$$

Multiplying out the product in Equation 4.2 and arguing exactly as before (Equations 1.5 to 1.9) we see that all previous closed graphs contribute to the expansion of $\langle \mu_k \mu_l \rangle$ in powers of $\tanh v$, but now the vertices k and l must contain an *odd number* of bonds rather than an even number in the expansion 1.9 of the partition function Z_N. Typical graphs are shown in Figure 6.12. A graph such as (a), with free ends and no intersections, corresponds to a *self-avoiding walk* from site k to l along the bonds of the lattice. No bond can occur more than once in a given graph, but self-intersections such as in (b) are allowed. The self-avoiding walk graphs of n bonds (or n steps) in a sense give the dominant contribution to the nth term in the $\tanh v$ expansion, but other graphs of course contribute.

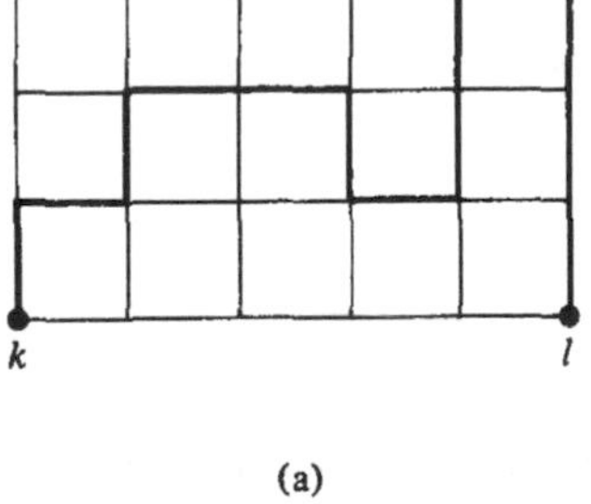

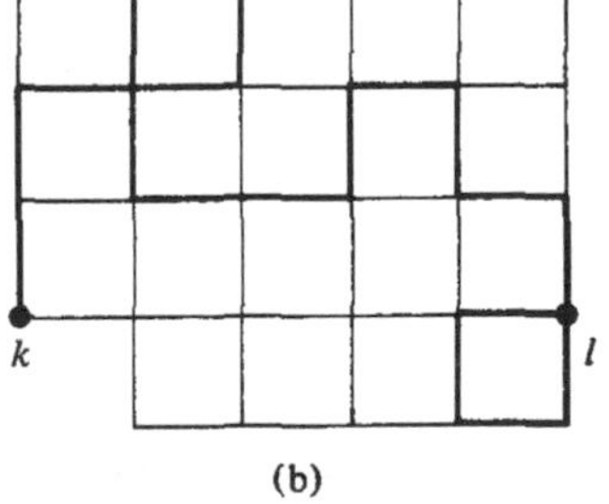

FIGURE 6.12. Typical graphs contributing to the high-temperature expansion of the pair-correlation function $\langle \mu_k \mu_l \rangle$.

The self-avoiding walk problem is of some independent mathematic interest, and also of interest as a simple model of a polymer chain. Typically one is interested in the number c_n of n-step self-avoiding walks and the mean-square end-to-end distance $\langle r_n^2 \rangle$ of n step walks. In one dimension the problem is trivial but in higher dimensions virtually nothing is known rigorously except that $\lim_{n\to\infty} n^{-1} \log c_n$ exists [Hammersley (1957)]. A great deal of configurational information in the form of exact enumeration of walks is known, however. Numerical analysis of such data (see the following section) and various heuristic arguments [e.g., Edwards (1965)] suggests that *if* $c_n \sim \mu^n n^g$ and $\langle r_n^2 \rangle \sim n^\theta$ as $n \to \infty$, then the "exponents" g and θ are $\frac{1}{3}$ and $\frac{3}{2}$ in two dimensions, and $\frac{1}{6}$ and $\frac{6}{5}$ in three dimensions, respectively. These numbers are very intriguing, especially the conjecture that

$$\theta = \begin{cases} \dfrac{6}{d+2} & \text{for dimensions } d = 1, 2, 3 \\ 1 & \text{for dimensions } d \geq 4. \end{cases} \tag{4.3}$$

(Note that $\theta = 1$ is the exact value for unrestricted random walks in any dimension.)

The behavior of the pair-correlation function for large separations of the two spins essentially determines the behavior of the zero-field susceptibility $\chi_0(T)$ in the neighborhood of the critical point. To see this we now derive the so-called fluctuation relation

$$\chi_0(T) = \beta\left(1 + \sum_{r \neq 0} \langle \mu_0 \mu_r \rangle\right) \qquad \text{for } T > T_c. \tag{4.4}$$

By definition the magnetization per spin $m_N(H, T)$ is given by

$$\begin{aligned} m_N(H, T) &= Z_N^{-1} \sum_{\{\mu\}} \left(\frac{1}{N} \sum_P \mu_P\right) \exp\left(\nu {\sum_{P,Q}}^{*} \mu_P \mu_Q + \beta H \sum_P \mu_P\right) \\ &= \left\langle \frac{1}{N} \sum_P \mu_P \right\rangle \\ &= \langle \mu_0 \rangle \end{aligned} \tag{4.5}$$

assuming translation invariance (0 denotes an arbitrary lattice point, e.g., the origin). The susceptibility is then given by (again assuming translation invariance)

$$\begin{aligned}\chi(H, T) &= \frac{\partial m_N}{\partial H} \\ &= \beta \sum_r (\langle \mu_0 \mu_r \rangle - \langle \mu_0 \rangle^2).\end{aligned} \tag{4.6}$$

Now, in zero field above the critical point the magnetization $\langle \mu_0 \rangle$ vanishes, so for an infinite lattice (since $\langle \mu_0^2 \rangle = 1$) we have, from Equation 4.6,

$$\chi_0(T) = \beta\left(1 + \sum_{r \neq 0} \langle \mu_0 \mu_r \rangle\right) \qquad \text{for } T > T_c. \tag{4.7}$$

Notice that this argument is not completely watertight because we have liberally interchanged limits $N \to \infty$ and $H \to 0$. Strictly speaking we should take the limit $N \to \infty$ *first* (so that the limit $N \to \infty$ of Equation 4.6 should first of all exist) and then take the limit $H \to 0+$. To derive Equation 4.7 we then must interchange a limit with an infinite sum. These operations appear reasonable, but as yet there is no proof of their validity.

Granted Equation 4.7, however, we can, following Fisher (1959b), deduce the high-temperature critical behavior of $\chi_0(T)$ from the known rigorous asymptotic behavior of the two-dimensional pair-correlation functions. Thus for temperatures above T_c we can write

$$\langle \mu_0 \mu_r \rangle \sim \exp(-\kappa r) w_r(T) \tag{4.8}$$

where $\kappa \sim c(T - T_c)$ as $T \to T_c +$ [Onsager (1944)]. Also, when $\mathbf{r} = (l, l)$ and $T = T_c$ [Kaufman (1949b)],

$$\langle \mu_0 \mu_r \rangle \sim A r^{-1/4} \qquad \text{as } r \to \infty. \tag{4.9}$$

For nondiagonal $\mathbf{r}$, Equation 4.9 is still valid, but A (and C) has a slight angular dependence. We therefore write

$$\langle \mu_0 \mu_r \rangle \sim A r^{-1/4} \exp[-C(T - T_c) r], \tag{4.10}$$

which should be valid as $r \to \infty$ and $T \to T_c +$. Subsequently we shall ignore the angular dependence of A and C. More detailed discussions of this point, and Equation 4.10, are given by Fisher (1959b), Kadanoff (1966), and Wu (1966). Replacing the sum by an integral (in polar coordinates) in Equation 4.7 we deduce from Equation 4.10 that

$$\begin{aligned}\chi_0(T) &\sim A\beta\pi \int_0^\infty r^{-1/4} \exp[-C(T - T_c)r] r \, dr \\ &= A\beta\pi C^{7/4}(T - T_c)^{-7/4} \int_0^\infty x^{3/4} \exp(-x) \, dx \\ &= D(T - T_c)^{-7/4},\end{aligned} \tag{4.11}$$

where

$$D = \beta A\pi C^{7/4}\Gamma(\tfrac{7}{4}), \tag{4.12}$$

and in the first step of Equation 4.11 we have changed variables to $x = C(T - T_c)r$.

Historically the $\frac{7}{4}$ power law, Equation 4.11, was first suggested by Domb and Sykes (1957) from analysis of high-temperature series expansions for $\chi_0(T)$. Numerical analysis of series expansion is the subject of the following section.

6-5 Numerical Analysis of the Three-dimensional Ising Model

We have remarked already that the basic combinatorial formula (1.9) is a high-temperature expansion and that one can compute $n(r)$, the number of closed graphs which can be constructed on the lattice, for values of r that are not too large. The degree of difficulty in computing $n(r)$ increases extremely rapidly with r, and even with modern high-speed computers it takes an enormous amount of time and effort to compute $n(r)$ for r larger than about 10 or so. Be that as it may, the development of series expansions for the partition function and other thermodynamic quantities begun some years ago by Wakefield (1951), Domb (1949), and others has now become, in the hands of Domb, Sykes, and co-workers at King's College, London, an art in itself. A large number of terms are known for various high- and low-temperature expansions for thermodynamic quantities such as specific heat, spontaneous magnetization, magnetic susceptibility, etc., on almost all conceivable lattices in two and three dimensions. The problem is to make the maximum use of this vast amount of information. What one would like to know, of course, is the location of the critical point and the behavior of various thermodynamic quantities in the neighborhood of that point for the three-dimensional Ising model.

Mathematically speaking, the simplest problem that could conceivably arise is the following: Given the first N terms $a_0, a_1, \ldots, a_{N-1}$ in the series expansion of a function $f(x)$,

$$f(x) = \sum_{k=0}^{\infty} a_k x^k \sim A(x - x_c)^{-\gamma} \qquad \text{as } x \to x_c, \tag{5.1}$$

can one deduce the location of the critical point x_c and the value of the critical exponent γ?

The answer simply is no, whatever the value of N! Consider, for example, the function

$$f(x) = \sum_{k=0}^{N-1} \left(\frac{x}{2}\right)^k + x^N(x-1)^{-1/2}. \tag{5.2}$$

For this function, the first N terms obviously give us no information whatever about the singularity at $x = 1$. The basic philosophy, then, or belief if you like, is that for Ising models the coefficients quickly settle down to the asymptotic behavior dictated by the singularity. Mathematically, this amounts to making some analytical assumptions about the function in question, and unfortunately we have no such information in three dimensions. We do, however, have information, in fact exact information, in two dimensions, so this provides a valuable testing ground for our numerical techniques. Thus, if it works in two dimensions, we hope that it also works in three dimensions.

On the historical side it is interesting to note that with no knowledge of the Onsager formula for the two-dimensional spontaneous magnetization m_0, Domb (1949) obtained series expansions for m_0 and using elementary extrapolation techniques (described below) suggested that $m_0 \sim (T_c - T)^{1/8}$ as $T \to T_c-$, which is absolutely correct (see Equation 5.22 of Chapter 5).

For reasons that we will state in a moment, the best (i.e., most well behaved) series to examine are the high-temperature zero-field susceptibility $[\chi_0(T)]$ series. In two dimensions the critical point is known exactly, but since the external magnetic field problem is unsolved, the exact formula for $\chi_0(T)$, or even its critical behavior, is not known rigorously. The "semirigorous argument" in the previous section, however, suggests that $\chi_0(T)$ is characterized by a high-temperature exponent $\gamma = \frac{7}{4}$ (see Equation 5.1) in two dimensions. We will see that the numerical methods support this value very strongly.

High-temperature expansions for $\chi_0(T)$ can be readily developed as discussed in the previous section, and in two and three dimensions one finds that all available terms (typically 20 in two dimensions and 15 in three dimensions) are positive, which means that the closest singularity to the origin must be on the real positive $\omega = \tanh(J/kT)$ axis, and consequently must correspond to the physical critical point (there will in general be other nonphysical singularities in the complex ω plane). The asymptotic behavior of coefficients in a series expansion of a function is determined by the closest singularity [the physical singularity in the case of $\chi_0(T)$], and the asymptotic behavior is naturally approached more rapidly if the closest singularity is sufficiently strong and well separated from other singularities. Fortunately, both these conditions appear to be fulfilled by the high-temperature susceptibility series.

To be a little more precise now, consider the high-temperature susceptibility expansion

$$\chi_0(T) = \sum_{n=0}^{\infty} a_n \omega^n \qquad \left[\omega = \tanh\left(\frac{J}{kT}\right)\right]. \tag{5.3}$$

For example, the triangular lattice series is given by

$$\begin{aligned}\chi_0(T) = 1 &+ 6\omega + 30\omega^2 + 138\omega^3 + 606\omega^4 + 2586\omega^5 \\ &+ 10{,}818\omega^6 + 44{,}574\omega^7 + 181{,}542\omega^8 + 732{,}678\omega^9 \\ &+ 2{,}935{,}218\omega^{10} + 11{,}687{,}202\omega^{11} + 46{,}296{,}210\omega^{12} + \cdots\end{aligned} \tag{5.4}$$

and the face-centered-cubic lattice series is given by

$$\begin{aligned}\chi_0(T) = 1 &+ 12\omega + 132\omega^2 + 1{,}404\omega^3 + 14{,}652\omega^4 + 151{,}116\omega^5 \\ &+ 1{,}546{,}332\omega^6 + 15{,}734{,}460\omega^7 + 159{,}425{,}580\omega^8 + \cdots.\end{aligned} \tag{5.5}$$

We now assume that in the neighborhood of the critical point $[\omega_c = \tanh(J/kT_c)]$, χ_0 has the form

$$\chi_0 \sim A\left(1 - \frac{\omega}{\omega_c}\right)^{-\gamma}. \tag{5.6}$$

If *all* coefficients a_n in Equation 5.3 are positive (and there is no strict proof), ω_c is the closest singularity to the origin. It then follows from the binomial theorem that the asymptotic form of the coefficients is given by

$$a_n \sim \frac{A}{\omega_c^n} \frac{(n + \gamma - 1)!}{n!(\gamma - 1)!} \qquad \text{as } n \to \infty. \tag{5.7}$$

The ratio of successive coefficients, therefore, has the asymptotic form

$$r_n = \frac{a_n}{a_{n-1}} \sim \omega_c^{-1}\left(1 + \frac{\gamma - 1}{n}\right) \qquad \text{as } n \to \infty. \tag{5.8}$$

A plot of r_n vs. n^{-1} should then asymptotically approach a straight line with intercept ω_c^{-1} and slope $\omega_c^{-1}(\gamma - 1)$. The results for the triangular and face-centered-cubic lattices are shown in Figure 6.13. For the triangular lattice the values of r_n computed from Equation 5.4 are

6.000, 5.000, 4.600, 4.391, 4.267, 4.183, 4.120, 4.073, 4.036, 4.006, 3.982, 3.961, ...,

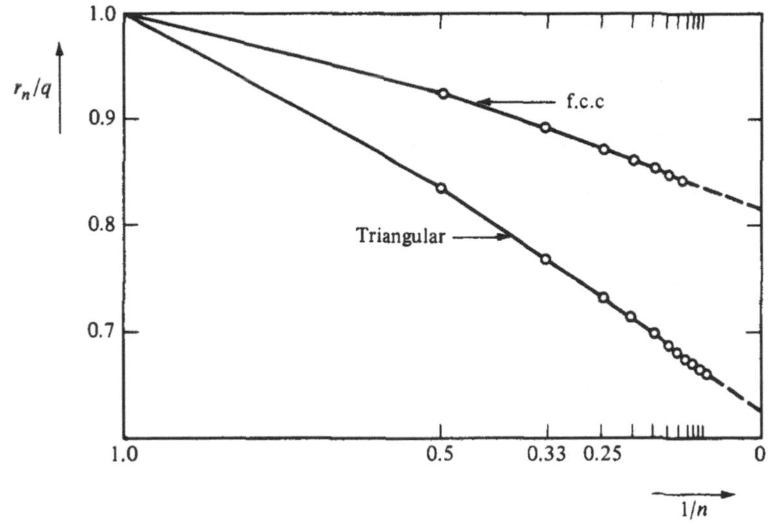

FIGURE 6.13. Ratios of successive coefficients r_n for the triangular ($q = 6$) and face-centered-cubic ($q = 12$) lattices.

which are decreasing smoothly, suggesting that

$$\omega_c^{-1} = 3.733 \pm 0.003. \tag{5.9}$$

The exact value (Equation 2.10) is

$$\omega_c^{-1} = 2 + \sqrt{3} = 3.73205\ldots, \tag{5.10}$$

so the extrapolated value is reasonably good.

The face-centered-cubic ratios settle down more rapidly than the triangular values, with successive r_n's given by

12.00, 11.00, 10.64, 10.45, 10.32, 10.22, 10.18, 10.13,

which suggest that

$$\omega_c^{-1} \approx 9.83. \tag{5.11}$$

The exact value is, of course, not known in this case.

Using the exact value of ω_c for the triangular lattice, we now compute successive values of

$$\gamma_n = 1 + n(r_n \omega_c - 1). \tag{5.12}$$

From Equation 5.8 $\lim_{n\to\infty} \gamma_n = \gamma$. Values for γ_n from Equation 5.4 are

0.609, 0.679, 0.698, 0.707, 0.717, 0.726, 0.728, 0.730, 0.733, 0.734, 0.736, 0.737, ...,

which are increasing smoothly, suggesting that

$$\gamma = 1.749 \pm 0.003, \tag{5.13}$$

and it is extremely tempting to conjecture that

$$\gamma = \tfrac{7}{4} \quad \text{exactly.} \tag{5.14}$$

Fisher's argument given in the previous section supports this value, but as we have already remarked, Equation 5.14 cannot yet be considered as exact.

A similar sequence is obtained for the face-centered-cubic γ, which suggests that

$$\gamma = 1.250 \pm 0.004, \tag{5.15}$$

and if one "believes," one would conclude that

$$\gamma = \tfrac{5}{4} \quad \text{exactly.} \tag{5.16}$$

The above method is usually referred to as the *ratio method.* Domb and Sykes (1961) used this method to obtain the results stated above.

So far we have only mentioned the "nice" series. The susceptibility series for the two-dimensional square and hexagonal lattices, and the three-dimensional simple cubic and body-centered cubic lattices (the "loose-packed lattices"), for example, have oscillating ratios r_n caused by the presence of a (antiferromagnetic) singularity at $-\omega_c$. The plots of r_n vs. n^{-1} in these cases are sawtoothed, so extrapolation is more difficult than for the triangular and face-centered-cubic lattices. The limit of r_n as $n \to \infty$ is still ω_c^{-1}, however, so what one does is either extrapolate the even and odd ratios separately or, better still, their arithmetic mean. The final results compare reasonably well with the triangular and face-centered-cubic results and suggest that $\frac{7}{4}$ and $\frac{5}{4}$ are the universal susceptibility exponents in two and three dimensions, respectively. (Recall that in two dimensions the specific heat and spontaneous magnetization have the same exact critical behavior for all lattices.)

The existence of a singularity at $-\omega_c$ for the loose-packed lattices is exact and can be deduced immediately from the basic combinational formula (1.9). Thus, as was shown in Section 6-1, only even powers of $\omega = \tanh \nu$ occur in the expansion of the square-lattice partition function. The same is true of all loose-packed lattices, and one concludes that if there is a singularity at $+\omega_c$ there must also be a singularity at $-\omega_c$. For an antiferromagnet $J < 0$, hence $\tanh v < 0$ for $T > 0$, so $-\omega_c$ corresponds to the antiferromagnetic singularity.

Although the occurrence of a pair of singularities at $\pm\omega_c$ disturbs the ratio method, the symmetry can be exploited [Guttmann *et al.* (1968)] if one makes one assumption about the form of $\chi_0(T)$—that the ferromagnetic singularity factors, i.e.,

$$\chi_0(T) = h(\omega)\left(1 - \frac{\omega}{\omega_c}\right)^{-\gamma}, \tag{5.17}$$

where $h(\omega)$ is analytic in the disc $|\omega| \leq \omega_c$ except at the point $\omega = -\omega_c$, where

$$h(\omega) \sim A\left(1 + \frac{\omega}{\omega_c}\right)^{-\alpha}. \tag{5.18}$$

The assumption 5.17 is not at all rigorous, but it appears from the series that it is probably true, or if not, it is very close to being true.

From Equations 5.17 and 5.18 we deduce that the function

$$h(\omega) = \left(1 - \frac{\omega}{\omega_c}\right)^{\gamma} \chi_0(T) \tag{5.19}$$

is singular at $-\omega_c$ and is analytic everywhere else in the region $|\omega| \leq \omega_c$. It follows that the coefficients of powers of ω/ω_c in the expansion of $h(\omega)$ alternate in sign and decrease in magnitude (provided α, Equation 5.18, is less than unity). The numerical procedure then is to choose trial values $\bar{\omega}_c$ and $\bar{\gamma}$ for ω_c and γ, generate the series for $h(\omega)$ (Equation 5.19), and see in fact if the coefficients alternate and decrease. In this way one obtains, with the aid of a computer, a set of possible values for ω_c and γ, which turn out to form a closed region in the $\bar{\omega}_c$, $\bar{\gamma}$ plane. A typical situation is shown schematically in Figure 6.14 [see Guttmann *et al.* (1968)].

We can carry the above procedure further by multiplying the series (Equation 5.19) by powers of $[1 + (\omega/\omega_c)]$ (for trial values of ω_c). This accelerates the rate of convergence of the series if the exact ω_c (and γ) are chosen, and amplifies any errors in the guessed values for ω_c and γ. Multiplying successively by $[1 + (\omega/\bar{\omega}_c)]^n$, $n = 1, 2, \ldots$, gives a sequence of smaller and smaller

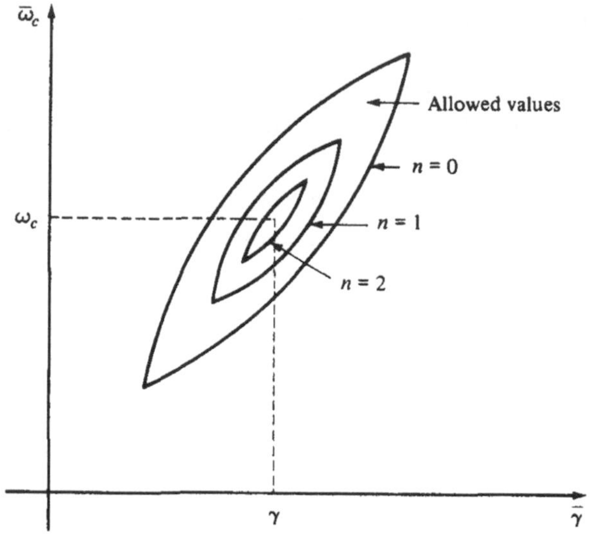

FIGURE 6.14. Possible values $\bar{\omega}_c$ and $\bar{\gamma}$ for the critical point ω_c and critical exponent γ obtained from Equation 5.19 by multiplying by powers of $(1 + \omega/\bar{\omega}_c)^n$, $n = 0, 1, 1, \ldots$.

regions as shown in Figure 6.14 and hence better estimates for ω_c and γ. One cannot, however, proceed indefinitely, since multiplying by powers of $[1 + (\omega/\bar{\omega}_c)]$ amplifies the effects of other singularities. This shows up in irregular behavior of the initial coefficients for $h(\omega)$ and makes it more difficult to judge if the series has settled down to the required alternating and decreasing behavior. It was found that with the available terms, $n = 2$ was about the largest reasonable value. For the square lattice, using the exact value $\omega_c^{-1} = 1 + \sqrt{2}$, the following results were obtained:

$$\begin{aligned} \gamma &= 1.750 \pm 0.005 \qquad (n = 0) \\ &\ 1.7497 \pm 0.0004 \qquad (n = 1) \\ &\ 1.7499 \pm 0.0002 \qquad (n = 2) \end{aligned} \tag{5.20}$$

where the value of n refers to the power of $[1 + (\omega/\bar{\omega}_c)]$, as described above. For the simple cubic lattice, where ω_c is not known, the results obtained were as follows:

$$\begin{aligned} \gamma &= 1.260 \pm 0.040 \qquad \omega_c = 0.21810 \pm 0.00090 \qquad (n = 0) \\ &\ 1.250 \pm 0.008 \qquad \ 0.21814 \pm 0.00020 \qquad (n = 1) \\ &\ 1.249 \pm 0.002 \qquad \ 0.21812 \pm 0.00004 \qquad (n = 2). \end{aligned} \tag{5.21}$$

Both results, Equations 5.20 and 5.21, suggest very strongly that $\gamma = \frac{7}{4}$ in two dimensions and $\gamma = \frac{5}{4}$ in three dimensions. One amusing result, assuming $\gamma = \frac{5}{4}$ exactly for the simple cubic lattice is that

$$\omega_c = 0.2181437 \pm 0.0000004 \qquad (n = 2), \tag{5.22}$$

which is obtained from 11 terms of the series!

For the close-packed lattices (triangular and face-centered-cubic) one does not have a singularity at $-\omega_c$. Such a singularity can, however, be introduced by considering the series for $\chi_0(T)\chi_0(-T)$. The above method can then be applied to this series and the results are almost as good as Equations 5.20 and 5.21. For example, for the face-centered-cubic lattice one obtains (from eight terms)

$$\begin{array}{lll} \gamma = 1.30 \pm 0.30 & \omega_c = 0.10179 \pm 0.00280 & (n = 0) \\ \quad 1.248 \pm 0.046 & \quad 0.10170 \pm 0.00018 & (n = 1) \\ \quad 1.248 \pm 0.011 & \quad 0.10175 \pm 0.00013 & (n = 2). \end{array} \tag{5.23}$$

Further results for the Ising model and self-avoiding walk problem can be found in Guttmann *et al.* (1968). So much for the well-behaved series.

Specific-heat series are difficult to analyze because the singularity is so weak (logarithmic in two dimensions and probably a small power in three dimensions). The above trick unfortunately does not work on specific-heat series because the (ferromagnetic) singularity almost surely does not factor (it certainly does not factor in two dimensions). Nevertheless, a straightforward analysis of the ratio of successive coefficients for the face-centered-cubic lattice suggests that the high-temperature specific-heat exponent α is probably $\frac{1}{8}$.

Low-temperature series in powers of $z = \exp(-2J/kT)$ (see Equation 2.4) are particularly difficult to analyze in three dimensions since with the exception of the diamond lattice, which we shall discuss in a moment, the series do not converge up to the physical singularity. For example, the simple cubic lattice appears to have its closest singularity on the negative z axis, so the ratio method gives us information about this unphysical singularity rather than the singularity of physical interest (i.e., the closest singularity on the positive z axis). The distribution of singularities in the complex z plane for the simple-cubic, body-centered-cubic, and face-centered-cubic lattices is shown in Figure 6.15 [see Guttman (1969a)].

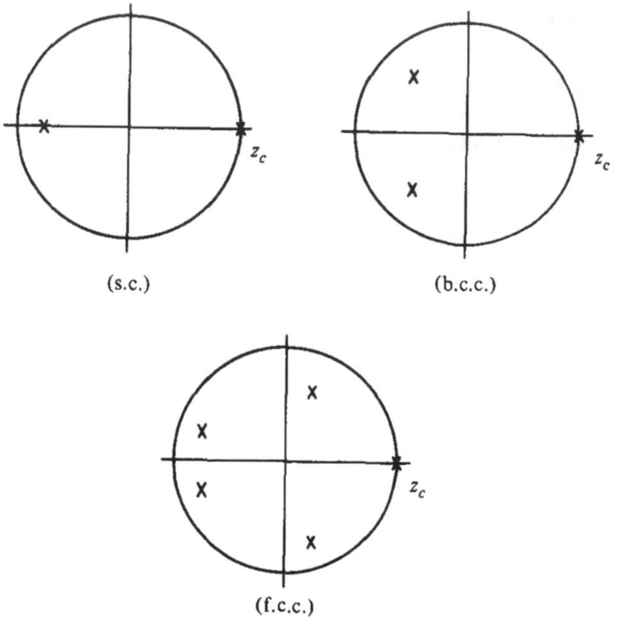

FIGURE 6.15. Distribution of singularities inside the disc $|z| \leq z_c$ for the simple cubic, body-centered-cubic, and face-centered-cubic lattices.

A number of methods have been applied to such series, one of the most successful to date being the method of *Padé approximants* [Baker (1961)]. The idea behind this method is to approximately analytically continue a series beyond its radius of convergence (i.e., the distance from the origin to the closest singularity) to examine the singularity of interest.

The $[N, M]$ Padé approximant of a function $f(z) = \sum_{n=0}^{\infty} a_n z^n$ is defined by

$$[N, M] = \frac{P_M(z)}{Q_N(z)}, \tag{5.24}$$

where

$$P_M(z) = \sum_{n=0}^{M} p_n z^n \qquad \text{and} \qquad Q_N(z) = \sum_{n=0}^{N} q_n z^n \tag{5.25}$$

are polynomials of degree M and N, respectively. Without loss of generality we can choose $q_0 = 1$. The remaining $M + N + 1$ coefficients $p_0, p_1, \ldots, p_M, q_1, \ldots, q_N$ are chosen so that the coefficients of the power-series expansion of $[N, M]$ agree with the first K (known) coefficients of the function $f(z)$. To determine the p_n's and q_n's we must then have $K = M + N + 1$.

If we denote the zeros of $Q_N(z)$ by z_i, $i = 1, 2, \ldots, N$ we can write (assuming $z_i \neq 0$, $i = 1, \ldots, N$)

$$[N, M] = \frac{P_M(z)}{\prod_{i=1}^{N}\left(1 - \dfrac{z}{z_i}\right)} \tag{5.26}$$

so that if $P_M(z_i) = R_i \neq 0$, and z_i is a simple root,

$$[N, M] \sim R_i\left(1 - \frac{z}{z_i}\right)^{-1} \qquad \text{as } z \to z_i. \tag{5.27}$$

Clearly, if the function $f(z)$ has only simple poles the Padé approximants will locate the positions of the singular points very accurately (in fact, exactly from $[N, M]$ if $f(z)$ has N poles).

To apply these ideas to the problem at hand, let us suppose that we have a function $f(z)$ which behaves as

$$f(z) \sim A\left(1 - \frac{z}{z_1}\right)^{-\gamma} \qquad \text{as } z \to z_1. \tag{5.28}$$

It follows that

$$\frac{d}{dz}[\log f(z)] \sim \frac{\gamma}{z_1 - z} \qquad \text{as } z \to z_1, \tag{5.29}$$

which has the required form for application of the Padé approximant method. Note that even if there are singularities closer to the origin than the one of interest [as there are for typical low-temperature Ising-model series in the variable $z = \exp(-2J/kT)$], the closest positive zero of $Q_N(z)$ to the origin in the $[N, M]$ Padé approximant to $(d/dz)[\log f(z)]$ should converge to the required singular point, and the residue of the pole at this point should, from Equation 5.29, converge to the critical exponent γ. Normally one considers $[N, N]$ Padé approximants, but others may be used (e.g., $[N \pm 1, N]$). Unfortunately, convergence properties of such Padé approximants are poorly understood, particularly in cases of interest, where taking the logarithmic derivative almost surely introduces branch points (this is, indeed, the case unless *all* singularities factor). In fact, some Padé approximants may not exist, which is also the case for some Ising-model series, so some care must be exercised with this method. Padé approximants do, however, work reasonably well on high-temperature susceptibility series, as shown in Table 6.1 (see

TABLE 6.1. $[N, N]$ Padé approximants for the high-temperature susceptibility-series critical points $[z_c = \tanh(J/kT_c)]$ and critical exponents (γ).

	Triangular		Simple Cubic	
N	z_c	γ	z_c	γ
1	0.2666667	1.706667	0.190476	0.870748
2	0.267054	1.712431	0.215107	1.204824
3	0.266714	1.705575	0.218967	1.280866
4	0.263923	1.751652	0.218151	1.250529
5	0.267950	1.749521	0.218183	1.251801
Exact	0.2679491 ...			

Equations 5.20 and 5.21), so one can have some confidence in applications of the method to low-temperature series. As examples, Padé sequences for the low-temperature spontaneous magnetization exponent β (see Table 6.1) obtained by Essam and Fisher (1963a) are

0.3106, 0.3032, 0.3120, 0.3218, 0.3133

for the simple cubic lattice, and

0.2824, 0.3045, 0.2970, 0.3029, 0.3161, 0.3089, 0.3067, 0.3115

for the face-centered-cubic lattice.

From these and other sequences Essam and Fisher suggested that

$$\beta = \tfrac{5}{16} \quad \text{exactly in three dimensions} \tag{5.30}$$

independently of the lattice. For the interested reader, a more detailed discussion of applications and convergence properties of Padé approximants can be found in Baker (1965).

The diamond lattice series, as we have mentioned, converge up to the physical singularity, so the simple ratio method can be applied. Essam and Sykes (1963b) applied this method to the logarithmic derivative of the spontaneous magnetization series and obtained the following sequence for β:

0.2974, 0.2995, 0.3031, 0.3059, 0.3071, 0.3081, 0.3088,

which supports the suggestion that $\beta = \frac{5}{16}$ (Equation 5.30).

The low-temperature susceptibility series are unfortunately not very smooth. The Padé sequences for the exponent γ' (see Table 4.1) are consequently rather irregular [Essam and Fisher (1963a)], and one cannot do much better than estimate $\gamma' = 1.28 \pm 0.05$. If one accepts the basic dogma in this field, however —that all critical exponents are rational numbers—the only "possible" values for γ' are $\frac{5}{4}$ or $1\frac{5}{16}$ (notice that the denominators of all exponents are then multiples of 4!). Recent Padé work on longer low-temperature susceptibility series, and on various indirect methods [Baker and Gaunt (1967) and Gaunt and Baker (1970)] have unfortunately failed to distinguish convincingly between the two "chosen values" for γ'. There are also insufficient terms available at present for the low-temperature susceptibility series on the diamond lattice to do any better than the Padé analysis on other series. Recently, however, it was noted [Guttmann and Thompson (1969b)] that all low-temperature series in three dimensions converge up to the critical point in the variable $x = 1 - \tanh(J/kT)$ (which is small when T is small). One pays the price, however, by having an uncomfortable number of singularities (e.g., eight for the face-centered-cubic lattice) close to the physical disc $|x| \leq x_c$ in the complex x plane. Nevertheless, one does rather well using only a mild extension of the ratio method on the logarithmic derivative of the various series. For example, one obtains the following sequence for the simple cubic exponent β:

0.2970, 0.2968, 0.3009, 0.3022, 0.3017, 0.3018, 0.3026, 0.3036, 0.3042, 0.3044, 0.3046, 0.3049, 0.3055, 0.3060,

which support the value $\beta = \frac{5}{16}$, and the following sequence for the simple cubic exponent γ':

1.238, 1.243, 1.251, 1.263, 1.281, 1.293, 1.299, 1.301, 1.303, 1.305, 1.308, 1.310, 1.311,

which is increasing, suggesting after one has played the game for a while that $\gamma' = 1\frac{5}{16}$ ($=1.3125$) exactly! Unfortunately, the sequences for other lattices are not quite as regular, and for these one cannot completely rule out the possibility that $\gamma' = \frac{5}{4}$. The high-temperature susceptibility exponent γ, as we have seen, is probably $\frac{5}{4}$ for all lattices. Various heuristic arguments (scaling, etc., discussed in Section 4-7) favor high temperature–low temperature, exponent symmetry (i.e., $\gamma = \gamma'$ and also $\alpha = \alpha'$ for the specific heat), so it would be very nice if γ' in fact turned out to be $\frac{5}{4}$. From the above sequence for γ', however, this may turn out not to be the case.

Low-temperature specific heat series are almost impossible to analyze except by indirect means which we shall not go into here. The conclusion is that the low-temperature specific-heat exponent α' is probably either $\frac{1}{16}$ or $\frac{1}{8}$, with the latter being slightly favored over the former.

Rushbrooke's inequality (Equation 7.1 of Chapter 4),

$$\alpha' + 2\beta + \gamma' \geq 2, \tag{5.31}$$

gives bounds for α' if one believes the estimates above for β and γ'. Thus, if we assume that $\beta = \frac{5}{16}$ and that $\frac{5}{4} \leq \gamma' \leq 1\frac{5}{16}$ we conclude from Equation 5.31 that

$$\alpha' \geq \tfrac{1}{16}. \tag{5.32}$$

From scaling and homogeneity arguments, which suggest that the inequality 5.31 is actually an equality, one concludes that

$$\alpha' = \tfrac{1}{16}, \quad \beta = \tfrac{5}{16}, \quad \gamma' = 1\tfrac{5}{16} \tag{5.33}$$

or

$$\alpha' = \tfrac{1}{8}, \quad \beta = \tfrac{5}{16}, \quad \gamma' = \tfrac{5}{4} \tag{5.34}$$

are the most likely values for the low-temperature exponents α', β, and γ', assuming that $\beta = \frac{5}{16}$ and that $\gamma' = 1\frac{5}{16}$ or $\frac{5}{4}$.

The final results for the two- and three-dimensional Ising models are summarized in Table 6.2, where for comparison we have included the classical and the experimental values (from Table 4.2). For completeness we have included estimates for δ, defined by (Table 4.1) $H \sim \mathrm{sgn}(m)|m|^{\delta}$ at $T = T_c$ as

TABLE 6.2. Estimates for two- and three-dimensional Ising-model exponents (α, α', and β are exact in two dimensions) compared with experimental values and the predictions of the classical theories.

Exponent	Two-dimensional	Three-dimensional	Classical	Experiment
α'	$0_{\log}$	$\frac{1}{8}$ or $\frac{1}{16}$	0_{disc}	$0_{\log} \lesssim \alpha' \lesssim 0.2$
α	$0_{\log}$	$\frac{1}{8}$	0_{disc}	$0_{\log} \lesssim \alpha \lesssim 0.2$
β	$\frac{1}{8}$	$\frac{5}{16}$	$\frac{1}{2}$	0.33
γ'	$\frac{7}{4}$	$\frac{5}{4}$ or $1\frac{5}{16}$	1	?
γ	$\frac{7}{4}$	$\frac{5}{4}$	1	1.3
δ	15	5 or $5\frac{1}{5}$	3	4.2

$H \to 0+$, where m is the magnetization and H is the magnetic field. The other exponents, α, α', β, γ, and γ', are, respectively, the high- and low-temperature specific-heat exponents, the spontaneous-magnetization exponent, and the high- and low-temperature susceptibility exponents. We have not included any error estimates since these are subjective and hence somewhat arbitrary. The two sets of values

$$\alpha' = \alpha = \tfrac{1}{8}, \quad \beta = \tfrac{5}{16}, \quad \gamma' = \gamma = \tfrac{5}{4}, \quad \delta = 5$$

and

$$\alpha' = \tfrac{1}{16}, \quad \alpha = \tfrac{1}{8}, \quad \beta = \tfrac{5}{16}, \quad \gamma' = 1\tfrac{5}{16}, \quad \gamma = \tfrac{5}{4}, \quad \delta = 5\tfrac{1}{5}$$

are, we believe, the most likely, with $\gamma = \frac{5}{4}$ and $\beta = \frac{5}{16}$ the most reliable. It is to be noted that these two sets of values satisfy the scaling laws (see Equations 7.22 of Chapter 4)

$$\begin{aligned} \alpha' + 2\beta + \gamma' &= 2, \\ \alpha' + \beta(1 + \delta) &= 2. \end{aligned} \tag{5.35}$$

Numerical estimates for other exponents, characterizing decay of correlations, etc., can be found in Fisher (1967) and Stanley (B1971).

There are finally two points to note:

1. The Ising-model exponents appear to depend only on dimensionality (and rather dramatically so) and not on the particular lattice (for α, α', and β this is exactly so for the two-dimensional Ising model). There is also a feeling among some people that the exponents do not depend on the range of interaction (as long as it is effectively finite). There is some numerical support for this feeling [Dalton and Wood (1969)], but at the moment it is not completely convincing.

2. The three-dimensional Ising-model exponents are not too different from the experimental values (at least they are closer to the experimental values than the classical exponents). Presumably the difference comes from the rigid and anistropic nature of the Ising model. (Numerical estimates for the more realistic isotopic Heisenberg model are in fact closer to the experimental values.) Hopefully if one could understand the two- and three-dimensional Ising-model exponents in a simple heuristic way, one would be well on the way to understanding the values that occur in nature.

PROBLEMS

1. Derive the zero-field partition function for the one-dimensional Ising model on (a) a closed (periodic) chain and (b) an open chain from the combinational formula (Equation 1.9)

$$Z_N = 2^N(\cosh v)^{Nq/2} \sum_{r=0}^{\infty} n(r)(\tanh v)^r.$$

2. Generalize the argument given in Section 6-2 to show that the partition function for a two-dimensional Ising model in an external magnetic field H can be written as

$$Z_N = \exp(\tfrac{1}{2}Nvq + N\beta H) \sum_{r,s=1}^{\infty} n_D(r, s)\exp(-2vr)\exp(-s\beta H),$$

where $n_D(r, s)$ is the number of closed graphs of r bonds on the dual lattice enclosing s points of the original lattice.

3. In the limit of an infinite square lattice (with N lattice points) the number of dimer configurations Φ is given by (Equations 32 and 33 of Appendix E)

$$\lim_{N\to\infty} \frac{2}{N} \log \Phi = (2\pi)^{-2} \iint_0^{2\pi} \log(4 - 2\cos\phi_1 - 2\cos\phi_2)\, d\phi_1\, d\phi_2$$
$$= F(2, 2),$$

where

$$F(x, y) = (2\pi)^{-2} \iint_0^{2\pi} \log(x + y - x\cos\phi_1 - y\cos\phi_2)\, d\phi_1\, d\phi_2.$$

Show that

(a) $F(x, 0) = \log(x/2), \qquad F(0, y) = \log(y/2);$

(b) $\dfrac{\partial F}{\partial x} = \dfrac{2}{\pi x} \tan^{-1}\left(\dfrac{x}{y}\right)^{1/2}.$

By expanding $\tan^{-1}(x/y)^{1/2}$ and integrating term by term, deduce that

$F(2, 2) = 4G/\pi,$

where $G = 1 - 3^{-2} + 5^{-2} - \cdots$ is Catalan's constant.

CHAPTER 7

Some Applications of the Ising Model to Biology

7-1 Introduction

The Ising and related models have been applied with some success to a number of biological systems. We shall discuss three examples here: hemoglobin, allosteric enzymes, and deoxyribonucleic acid (DNA). These examples are by no means exhaustive, but they illustrate how general lattice combinatoric problems and their methods of solution may be applied to biological problems.

The common feature in the examples we shall consider here is "cooperativity," which is playing a role of increasing importance in biology these days. Hemoglobin, for example, which is the oxygen carrier in red blood cells, has four distinct binding sites for oxygen with apparent interactions between the sites. In other words, if a molecule of oxygen is bound to a hemoglobin molecule, it will be more likely than not to bind other oxygen molecules. This cooperative interaction between the binding sites reflects itself in the saturation curve shown in Figure 7.1, where the percentage of oxygenated hemoglobin is plotted as a function of the partial pressure of oxygen (or equivalently the concentration of oxygen).

It is clear from the S, or sigmoid, shape of the hemoglobin saturation curve that binding of a few molecules of oxygen favors the binding of more and

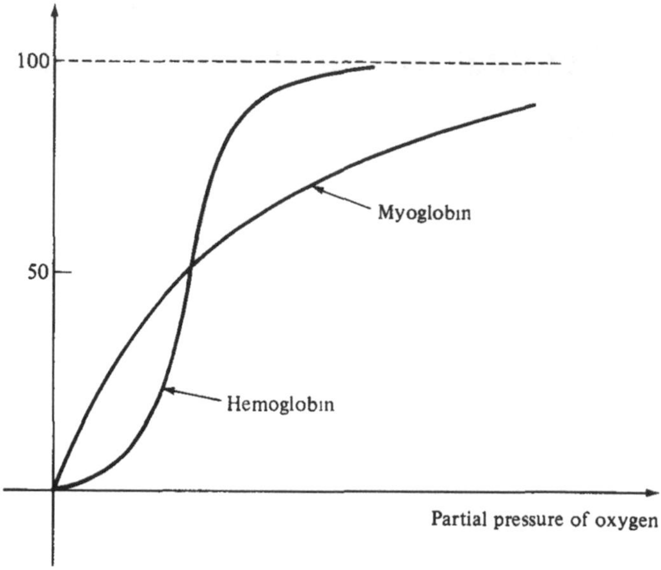

FIGURE 7.1. Percentage of oxygenated hemoglobin and myoblogin as a function of the partial pressure of oxygen.

that binding occurs best at high oxygen concentrations. This naturally makes the hemoglobin molecule a very efficient little machine, since in the lungs where the oxygen concentration is high binding is easy, and, as hemoglobin travels through the body to places with low oxygen concentration, the oxygen leaves the molecule rather easily.

In Figure 7.1 the saturation curve for hemoglobin is compared with the corresponding curve for myoglobin, the oxygen-bearing molecule found in muscle tissue, which has only one binding site for oxygen and hence no chance of displaying any cooperative effects.

A similar situation holds for enzymes, which are catalysts for biochemical reactions. The agent undergoing a particular reaction is usually referred to as the substrate, and the reaction of the substrate, catalyzed by an enzyme, proceeds by substrate molecules occupying distinct binding sites on the enzyme. For a "classical enzyme," the initial reaction rate, which is assumed to be proportional to the number of occupied sites on the enzyme, as a function of substrate concentration, has the same form as the myoglobin saturation curve, showing that there is no apparent interaction between the (possibly many) distinct binding sites. A number of enzymes, the "allosteric enzymes," were recently found, however, to have an S-shaped initial reaction-rate curve reflecting some degree of cooperativity among the binding sites.

Finally, DNA also has an S-shaped denaturation or melting curve; i.e., the fraction of broken bonds as a function of temperature has the same form as the hemoglobin saturation curve. If one accepts the Watson–Crick model for DNA, which pictures the molecule as a double helix with hydrogen bonds connecting the two strands, denaturation occurs through the breaking of the hydrogen bonds as the temperature is increased. The denaturation process is then often referred to as the helix-coil transition, and the fraction of broken bonds as a function of temperature gives the "melting curve." The S shape of the melting curves presumably means that the bond-breaking process is cooperative rather than random.

To set the stage for a discussion of the cooperative processes described above we present in the next section a model for the myoglobin and classical enzyme noncooperative processes. In the following sections we shall discuss the structure and function of hemoglobin, allosteric enzymes, and DNA in more detail and show how the random or independent model in Section 7-2 can be modified to take cooperativity into account.

For some background, the interested reader is referred to Perutz (1964) for hemoglobin, Changeux (1965) for enzymes, and Watson (B1968) for DNA.

7-2 Myoglobin and Classical Enzymes

Consider an enzyme with n noninteracting or independent binding sites for substrate ($n = 1$ corresponds to myoglobin and in this case the "substrate" is oxygen). We number the sites by an index $i = 1, 2, \ldots, n$ and associate with each site a parameter μ_i which takes two values, $+1$ if the ith site is occupied by substrate and -1 if the ith site is unoccupied by substrate. (It would perhaps be better to consider lattice gas variables $t_i = 1$ or 0 instead of $\mu_i = +1$ or -1, respectively, but we will see in the following sections that the relation to the Ising model is made easier by starting with the Ising magnet variables $\mu_i = \pm 1$.)

A configuration of the molecule is specified by the values of $\mu_1, \mu_2, \ldots, \mu_n$. In other words, a particular configuration specifies which sites are occupied and which sites are unoccupied. Since there are two states for each site, there are 2^n possible configurations. Henceforth we shall denote a configuration by

$$\{\mu\} = (\mu_1, \mu_2, \ldots, \mu_n). \tag{2.1}$$

The quantity we wish to calculate is the average number of occupied sites on the enzyme. In a *particular configuration* $\{\mu\}$, the number $N\{\mu\}$ of occupied sites is given by

$$N\{\mu\} = \sum_{i=1}^{n} \tfrac{1}{2}(1 + \mu_i) \tag{2.2}$$

since by definition $\frac{1}{2}(1 + \mu_i)$ is 1 when the ith site is occupied and zero when the ith site is unoccupied. The average number of occupied sites N is then defined by

$$N = \sum_{\{\mu\}} N\{\mu\}P\{\mu\}, \tag{2.3}$$

where the sum is over all (2^n) configurations $\mu_1 = \pm 1, \ldots, \mu_n = \pm 1$ and $P\{\mu\}$ is the probability of the configuration $\{\mu\}$.

So far the discussion has been completely general. To proceed further we must specify the probability distribution $P\{\mu\}$. In the case of independent binding sites we define $p(+1)$ to be the probability that a site is occupied and $p(-1)$ to be the probability that a site is unoccupied. Since a site is either occupied or unoccupied we must have

$$p(+1) + p(-1) = 1. \tag{2.4}$$

To simplify the algebra we now define quantities C and J by

$$p(+1) = Ce^{J}, \qquad p(-1) = Ce^{-J}, \tag{2.5}$$

where, from Equation 2.4,

$$C = (2 \cosh J)^{-1}. \tag{2.6}$$

Equation 2.5 can then be written simply as

$$p(\mu_i) = (2 \cosh J)^{-1} \exp(J\mu_i), \tag{2.7}$$

which gives the probability distribution for the ith site.

Now in the case of independent binding sites the probability distribution for the whole molecule $P\{\mu\}$ is simply the product of the probability distributions for each site, i.e.,

$$P\{\mu\} = \prod_{j=1}^{n} p(\mu_j\}. \tag{2.8}$$

We are now in a position to calculate the average number N of occupied sites. From Equations 2.2, 2.3, 2.7, and 2.8 we have

$$N = \tfrac{1}{2}\sum_{i=1}^{n} \sum_{\{\mu\}} (1 + \mu_i) \prod_{j=1}^{n} p(\mu_j), \tag{2.9}$$

and from the normalization condition 2.4,

$$\begin{aligned} \sum_{\{\mu\}} \prod_{j=1}^{n} p(\mu_j) &= \sum_{\mu_1=\pm 1} p(\mu_1) \sum_{\mu_2=\pm 1} p(\mu_2) \cdots \sum_{\mu_n=\pm 1} p(\mu_n) \\ &= 1. \end{aligned} \tag{2.10}$$

Also, after summing over μ_j, $j \neq i$, in Equation 2.9, we obtain

$$\begin{aligned} \sum_{\{\mu\}} \mu_i \prod_{j=1}^{n} p(\mu_j) &= \sum_{\mu_i=\pm 1} \mu_i \, p(\mu_i) \\ &= p(+1) - p(-1) \\ &= \tanh J, \end{aligned} \tag{2.11}$$

where in the last step we have used Equations 2.5 and 2.6. Combining Equations 2.9, 2.10, and 2.11 we then have

$$N = \frac{n}{2}(1 + \tanh J). \tag{2.12}$$

Defining α by

$$\alpha = \exp(2J) \tag{2.13}$$

we can write Equation 2.12 in the form

$$f(\alpha) = \frac{N}{n} = \frac{\alpha}{1 + \alpha}, \tag{2.14}$$

which is the classical Michaelis–Henri equation for enzyme kinetics if α is interpreted as the concentration of substrate. From Equations 2.5 and 2.13, α is the ratio of the probability that a site is occupied to the probability that a site is unoccupied, so it seems reasonable to interpret α as a measure of the concentration of substrate.

The Michaelis–Henri equation (2.14) fits the myoglobin saturation curve (Figure 7.1) and the initial reaction-rate curve for classical enzymes extremely well. Note that the "units" in Equation 2.14 are chosen so that half-saturation occurs at concentration unity.

FIGURE 7.2. Schematic illustration of the hemoglobin molecule showing the four hemes (hatched) and the associated α (white) and β (black) amino acid chains.

7-3 Hemoglobin

The hemoglobin molecule is huge, with a molecular weight of about 64,500. It is composed of about 10,000 atoms of hydrogen, carbon, nitrogen, oxygen, sulfur, and four atoms of iron which lie at the center of a group of atoms that form the pigment heme. The heme gives blood its red color, and each heme constitutes a binding site for an oxygen molecule.

The structure of hemoglobin, shown schematically in Figure 7.2, was first unraveled by Perutz et al. (1960) after 23 years or so of painstaking work. Their primary tool was X-ray diffraction, which came into its own in the 1950s as a means for studying the structure of large molecules. Using X-ray diffraction techniques, for example, Kendrew found the structure of myoglobin [see Perutz (1964)] and Wilkins's X-ray pictures of DNA lead to the discovery of the Watson–Crick double-helix model for DNA [see Watson (B1968)]. All these remarkable events occurred in London and Cambridge (England) in the 1950s.

As shown in Figure 7.2, hemoglobin consists of four (hatched) binding sites, or hemes, and, in normal form, two types of amino acid chains, the α and β chains, shown as white and black, respectively, in Figure 7.2. The unlike chains are rather strongly coupled to one another, whereas the like chains have virtually no contacts with one another. To a good approximation the hemes are at the vertices of a regular tetrahedron, shown schematically in Figure 7.3 with the associated amino acid chains. The slight asymmetry of the real molecule reflects itself in the different number of contacts the β chains have with the α chains; in one case 34 residues participate and in the other case 19 residues participate. We shall assume for simplicity that the *degree of cooperativity or interaction* among hemes coupled by different chains is symmetric. (The nonsymmetric case can also be handled with the present formalism.) We shall also assume that interactions among the hemes are transmitted solely through the α-β contacts. In other words, from Figure 7.3, heme 1 interacts with hemes 2 and 4 but not with 3, heme 2 interacts with hemes 1 and 3 but not with 4, and so forth. *This means that we can represent the hemoglobin molecule by a ring of four sites with nearest-neighbor sites only interacting.* To be a little more general, let us consider now a ring of n sites as shown in Figure 7.4. The problem is to calculate the average number of

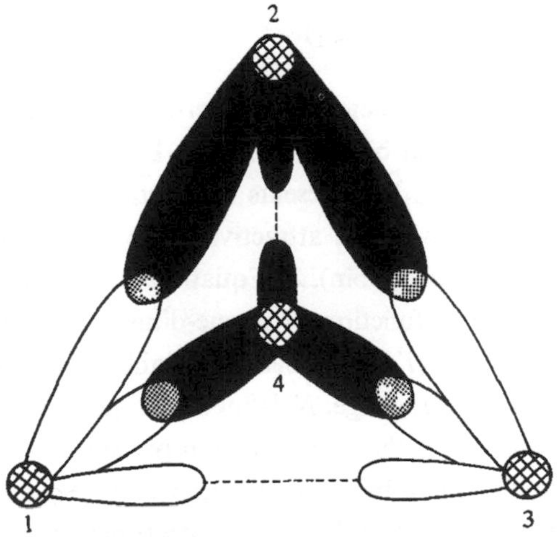

FIGURE 7.3. Simplified model for hemoglobin. The four hemes are approximately situated on the vertices of a regular tetrahedron. Unlike chains are rather strongly coupled, whereas like chains have virtually no contacts with one another.

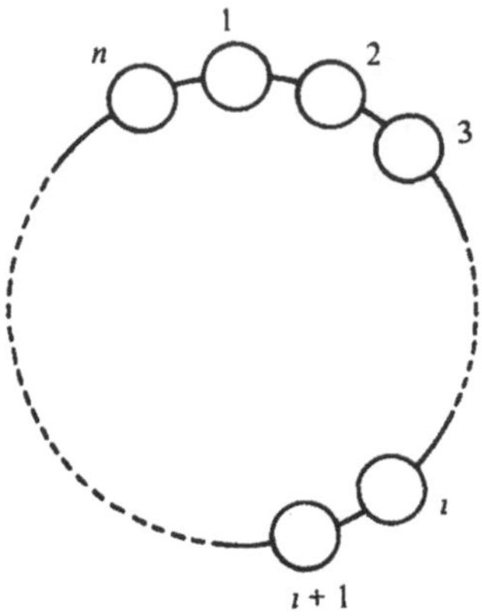

FIGURE 7.4. Nearest-neighbor interaction model for a molecule consisting of n binding sites.

occupied sites defined by Equation 2.3. Assuming that only nearest-neighbor sites interact, we can take the probability distribution $P\{\mu\}$ to be of the form

$$P\{\mu\} = Z^{-1} \prod_{i=1}^{n} \exp(J\mu_i)\exp(U\mu_i \mu_{i+1}), \tag{3.1}$$

where Z is such that $P\{\mu\}$ is properly normalized $\left(\sum_{\{\mu\}} P\{\mu\} = 1\right)$, i.e.,

$$Z = \sum_{\{\mu\}} \prod_{i=1}^{n} \exp(J\mu_i)\exp(U\mu_i \mu_{i+1}) \tag{3.2}$$

and $\mu_{n+1} = \mu_1$. The first factor in the product (Equation 3.1), $\exp(J\mu_i)$, represents the independent binding term (see Equations 2.7 and 2.8) and the second term, $\exp(U\mu_i \mu_{i+1})$, represents interactions or correlations between nearest-neighbor sites, which is attractive when $U > 0$ and repulsive when $U < 0$ (so $U > 0$ for hemoglobin). Z (Equation 3.2) will be recognized immediately as the partition function of a one-dimensional Ising magnet in an external magnetic field kTJ, with nearest-neighbor coupling constant equal to $-kTU$. In magnetic language, N defined by Equations 2.2 and 2.3 is equal to $n(1 + m)/2$, where m is the magnetization per spin.

One point which should be stressed is that $P\{\mu\}$ *(Equation 3.1) is the most general probability distribution describing nearest-neighbor correlations.* [Since $\mu_i^2 = 1$ and $(\mu_i \mu_{i+1})^2 = 1$, the most general functions of μ_i and of $\mu_i \mu_{i+1}$ are linear functions, and as was shown in Section 6-1, linear functions of a given product of μ's can always be expressed as exponentials, and vice versa.]

To evaluate the average number of occupied sites N (Equation 2.3), we note first of all from Equations 2.2, 3.1, and 3.2 that

$$
\begin{aligned}
N &= \sum_{\{\mu\}} \left[\frac{1}{2} \sum_{i=1}^{n} (1 + \mu_i) \right] P\{\mu\} \\
&= \frac{n}{2} + \frac{1}{2} \sum_{\{\mu\}} \left(\sum_{i=1}^{n} \mu_i \right) P\{\mu\} \qquad (3.3) \\
&= \frac{n}{2} + \frac{1}{2} \frac{\partial}{\partial J} (\log Z).
\end{aligned}
$$

Now, in Section 5-3 it was shown that

$$Z = \lambda_1^n + \lambda_2^n \qquad \text{(Equation 3.12 of Chapter 5)} \qquad (3.4)$$

where λ_1 and λ_2 are the eigenvalues of the transfer matrix (Equation 3.10 of Chapter 5)

$$\mathbf{L} = \begin{pmatrix} \exp(U + J) & \exp(-U) \\ \exp(-U) & \exp(U - J) \end{pmatrix}, \qquad (3.5)$$

i.e. (Equation 3.14 of Chapter 5),

$$\left.\begin{matrix} \lambda_1 \\ \lambda_2 \end{matrix}\right\} = e^U \cosh J \pm (e^{-2U} + e^{2U} \sinh^2 J)^{1/2}. \qquad (3.6)$$

Equations 3.3 and 3.4 then give

$$N = \frac{n}{2} \left[1 + (\lambda_1^n + \lambda_2^n)^{-1} \left(\lambda_1^{n-1} \frac{\partial \lambda_1}{\partial J} + \lambda_2^{n-1} \frac{\partial \lambda_2}{\partial J} \right) \right], \qquad (3.7)$$

and it is straightforward, but rather tedious, to show from Equation 3.6 that in terms of the interaction parameter U and the concentration $\alpha = \exp(2J)$ (Equation 2.13) the average fraction of occupied sites is given by

$$
\begin{aligned}
f(\alpha) &= \frac{N}{n} \\
&= \frac{\alpha\{(1 + \alpha + \delta)^{n-1}[1 + (2e^{-4U} + \alpha - 1)/\delta] + (1 + \alpha - \delta)^{n-1}[1 - (2e^{-4U} + \alpha - 1)/\delta]\}}{(1 + \alpha + \delta)^n + (1 + \alpha - \delta)^n}, \qquad (3.8)
\end{aligned}
$$

where

$$\delta = [(\alpha - 1)^2 + 4\alpha \exp(-4U)]^{1/2}. \qquad (3.9)$$

It is to be noted that in the limit $U = 0$, Equation 3.8 reduces to the classical Michaelis–Henri equation (2.14). In the limit $U \to \infty$, $\delta = |\alpha - 1|$, and Equation 3.8 reduces to the Hill equation

$$f(\alpha) = \frac{\alpha^n}{1 + \alpha^n}, \tag{3.10}$$

after Hill, who first attempted to fit hemoglobin saturation curves to this form (the fit is actually reasonably good for $n \approx 2.6$).

For hemoglobin, $n = 4$ and Equation 3.8 takes the simple form

$$f(\alpha) = \frac{\alpha[K + (2K + K^2)\alpha + 3K\alpha^2 + \alpha^3]}{1 + 4K\alpha + (4K + 2K^2)\alpha^2 + 4K\alpha^3 + \alpha^4}, \tag{3.11}$$

where

$$K = \exp(-4U). \tag{3.12}$$

Apart from trivial changes in the scale of concentration, Equation 3.11 is identical with Pauling's model [Pauling (1935)] and is a special case of the square model of Koshland et al. (1966).

In Equation 3.11 there is only one adjustable parameter, K, which is determined by fitting to one experimental point. The resulting theoretical curve for the choice $K = 0.11$ (i.e., from Equation 3.12, $U = 0.55$), is compared with experiment in Figure 7.5. The agreement is reasonably good, but there appears to be a slight systematic deviation from the theoretical curve at high oxygen concentrations. This is probably due to the asymmetric nature of the molecule. As mentioned above, this can be taken into account, specifically in the model, by taking two coupling constants U_1 and U_2 for the two different contacts between unlike chains. With one extra parameter, however, a perfect fit to experiment is almost assured. In fact, a number of two-parameter models, particularly the Monod–Wyman–Changeux model discussed in the following section, give remarkably good results. What is perhaps surprising is the fit obtained using only a one-parameter model.

One interesting feature of hemoglobin saturation curves is that the degree of cooperativity appears to be independent of pH and temperature. In other words, identical saturation curves are obtained over a range of pH and temperature if the concentration scale is chosen (for example) so that half-saturation always occurs at concentration unity. This means in particular that our model parameter $U = 0.55$ is independent of pH and temperature (J, of course, does depend on pH and temperature).

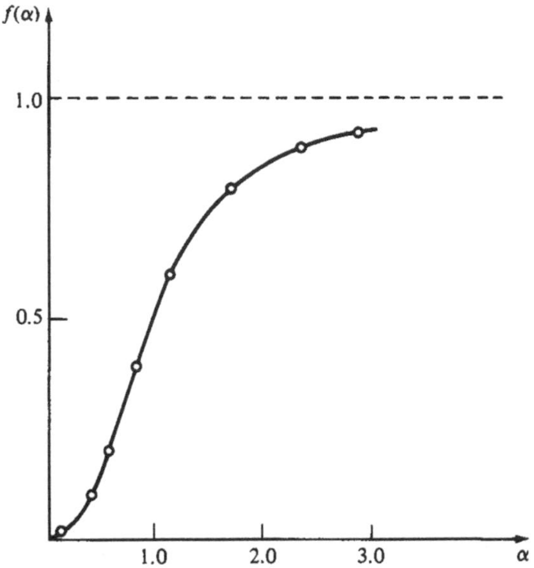

FIGURE 7.5. Fraction of oxygenated hemoglobin $f(\alpha)$ as a function of oxygen concentration α in units such that half-saturation occurs at concentration unity. The solid curve is obtained from Equation 3.11 with $U = 0.55$ and the circles are experimental values taken from Changeux (1965).

7-4 Monod–Wyman–Changeux Model for Hemoglobin and Allosteric Enzymes

Hemoglobin is considered as a prototype for the recently discovered allosteric enzymes, which are characterized by sigmoid-shaped initial reaction-rate (or saturation) curves, rather than by curves of the Michaelis–Henri type for classical enzymes (Figure 7.1). Hemoglobin is, of course, not an enzyme, but it does have an S-shaped saturation curve, and its does have distinct binding sites for "substrate" (i.e., oxygen). Unfortunately, almost nothing is known about the structure of allosteric enzymes except that they have distinct binding sites for substrate.

Typically, allosteric enzymes are also regulatory enzymes. The enzyme aspartate transcarbamylase (ATCase), for example, with aspartate as substrate catalyses a reaction with citosine triphosphate (CTP) as an end product. With CTP present, the initial aspartate reaction is inhibited. In other words, CTP controls or regulates the rate of its own production. Such regulatory or feed-back behavior appears to be extremely important in body chemistry and is thought to be caused by cooperative interactions (or "allosteric effects") among the binding sites on the enzyme [Changeux (1965)].

A model for such behavior was proposed in 1965 by Monod, Wyman, and Changeux (1965). We shall discuss this (MWC) model here before discussing some generalizations of the Ising model as models for allosteric enzymes in the following section.

The MWC model assumes first of all that the *whole molecule* can exist in one of two conformational states (i.e., states with different "shapes") denoted by R (for relaxed) and T (for tense), respectively, *independently of whether substrate is bound to the enzyme or not*. X-ray data for hemoglobin show that the molecule does undergo conformational changes when oxygen binds to the molecule, but not otherwise; so it would seem that the first assumption in the MWC model is rather unrealistic. The second assumption is that binding in each of the two conformational states is an independent process. That is, there are no interactions or correlations between the binding sites. This also appears to be a little unrealistic in view of our discussion in the previous section.

Let us consider first an enzyme in the presence of substrate only. In the present language the MWC is then very simple. We introduce an additional parameter σ which takes values $+1$ if the molecule is in an R state and -1 if the molecule is in a T state. A configuration is then specified by

$$\{\sigma; \mu\} = (\sigma; \mu_1, \mu_2, \ldots, \mu_n), \tag{4.1}$$

the σ to tell which state the molecule is in and the $\mu_i = \pm 1$, $i = 1, 2, \ldots, n$, as before, to tell which sites are occupied and which sites are unoccupied. Since binding is assumed to occur independently in each state (R or T) the probability distribution is a sum of two independent distributions (Equation 2.8), i.e.,

$$P\{\sigma; \mu\} = Z^{-1}\left[\tfrac{1}{2}(1+\sigma)\prod_{i=1}^{n} e^{J_1(1+\mu_i)} + \frac{L}{2}(1-\sigma)\prod_{i=1}^{n} e^{J_2(1+\mu_i)}\right], \tag{4.2}$$

where for normalization

$$\begin{aligned} Z &= \sum_{\{\sigma;\mu\}}\left[\frac{1}{2}(1+\sigma)\prod_{i=1}^{n} e^{J_1(1+\mu_i)} + \frac{L}{2}(1-\sigma)\prod_{i=1}^{n} e^{J_2(1+\mu_i)}\right] \\ &= \sum_{\{\mu\}}\prod_{i=1}^{n} e^{J_1(1+\mu_i)} + L\sum_{\{\mu\}}\prod_{i=1}^{n} e^{J_2(1+\mu_i)} \\ &= (1+\alpha)^n + L(1+c\alpha)^n. \end{aligned} \tag{4.3}$$

$$\alpha = \exp(2J_1) \tag{4.4}$$

is the ratio of the probability that a site is occupied to the probability that a site is unoccupied in the R state and

$$c\alpha = \exp(2J_2) \tag{4.5}$$

is the ratio of the probability that a site is occupied to the probability that a site is unoccupied in the T state. α, as before, is assumed to be a measure of the concentration of substrate, and L in Equation 4.2 is the "allosteric constant," i.e., the ratio of the probability of the T state to the probability of the R state in the absence of substrate (i.e., all $\mu_i = -1$).

The average number of occupied sites N is easily found to be given by (see Problem 1)

$$\begin{aligned}\frac{N}{n} &= \frac{1}{2n}\sum_{i=1}^{n}\left[\sum_{\{\sigma;\,\mu\}}(1+\mu_i)P\{\sigma;\mu\}\right]\\ &= \frac{\alpha(1+\alpha)^{n-1} + Lc\alpha(1+c\alpha)^{n-1}}{(1+\alpha)^n + L(1+c\alpha)^n}.\end{aligned} \tag{4.6}$$

The similarity in form of Equations 4.6 and 3.8 should be noted.

In the MWC model an S-shaped saturation curve is obtained, for example, by taking L small and c large, which is to say that the number of T molecules is small compared with the number of R molecules, but the probability of binding to a T molecule is large compared with the probability of binding to an R molecule. It is this competitive effect that produces in a somewhat artificial way the characteristic S-shaped saturation curve.

Equation 4.6 fits hemoglobin and allosteric enzyme data rather well [Monod et al. (1965)], which is perhaps not surprising in view of the fact that there are two adjustable parameters (L and c). As we have remarked, the basic assumptions of the model seem to be rather unrealistic since it is likely that there are interactions between binding sites, and that conformational changes of the molecule and binding of substrate to the molecule are not independent of one another. Notice also that the model takes no account of structure. For hemoglobin this is a definite defect in the model, but for allosteric enzymes, where almost nothing is known about structure, it may be an advantage. By considering more sophisticated models, however, it may be possible to deduce some structural information (but probably not a great deal) from the saturation curves. Some work along these lines is discussed in the following section.

In conclusion we remark that the MWC model above can be generalized to take various modifiers (e.g., CTP for ATCase) into account (see Problem 2). These generalized models also fit the experimental data rather well.

7-5 Decorated Ising Models for Allosteric Enzymes

There are 30 or so known allosteric enzymes, but unfortunately almost nothing is known about their structure, even their quaternary structure, i.e., the number of subunits or binding sites. The most studied allosteric enzyme has been aspartate transcarbamylase (ATCase). Early experiments on ATCase [Gerhart and Pardee (1963) and Gerhart and Schachman (1965)] demonstrated that there were separate distinct binding sites for substrate and for modifiers (e.g., the inhibitor CTP). This appears to be still the case. The number of binding sites in each case, however, is a little in doubt at the moment. The original experiments suggested that there were eight binding sites for substrate (catalytic sites) and eight binding sites for modifiers (regulatory sites). More recent experiments by the original group [Changeux and Schachman (1968a), Changeux and Rubin (1968b) and Gerhart and Schachman (1968)] suggest four regulatory and four catalytic sites, whereas an independent group [Weber (1968)] suggests six catalytic sites and six regulatory sites. Let us suppose then that there are $n(4 \leq n \leq 8)$ catalytic sites and $m(4 \leq m \leq 8)$ regulatory sites on the enzyme.

A possible model for ATCase in the absence of modifiers is the model discussed in Section 7-3 with probability distribution given by Equation 3.1 and average number of occupied (catalytic) sites given by Equation 3.8.

For the model to be reasonably close to the truth we need the molecule to have a one-dimensional interaction structure. By this we mean that contacts between subunits can be traced along a one-dimensional path, as for hemoglobin. The molecule will, of course, have a three-dimensional shape, as hemoglobin does, if for no other reason than the molecule will want to occupy the minimum volume.

On general grounds one might argue that all biological molecules have essentially a one-dimensional interaction structure. If this were not so, it would be virtually impossible to make such molecules either naturally or artificially by stringing together components in a definite sequence, which is the way it is done artificially and is probably the way nature does it as well.

If the one-dimensional interaction structure is granted, and if it is assumed that only nearest-neighbor subunits interact, the probability distribution (3.1) for configurations of the molecule is the only possibility.

The fact that the number of subunits is not known precisely is of little concern, since for the model the fraction of occupied sites given by Equation 3.8 is rather insensitive to n for $n \geq 4$ or so. Thus in Equation 3.8 if we divide top and bottom by $(1 + \alpha + \delta)^n$ and allow n to approach infinity we obtain, since

$$\lim_{n\to\infty} \left(\frac{1 + \alpha - \delta}{1 + \alpha + \delta}\right)^n = 0,$$

$$\lim_{n\to\infty} \frac{N}{n} = \frac{\alpha(\delta + \alpha - 1 + 2e^{-4U})}{\delta(1 + \alpha + \delta)}. \tag{5.1}$$

The error committed in approximating Equation 3.8 by Equation 5.1 for finite n is then exponentially small in n and hence constitutes a negligible error for n larger than about 4.

In Figure 7.6 we have compared a theoretical curve obtained from Equation 3.8 with $n = 8$, with experimental values obtained for ATCase in the

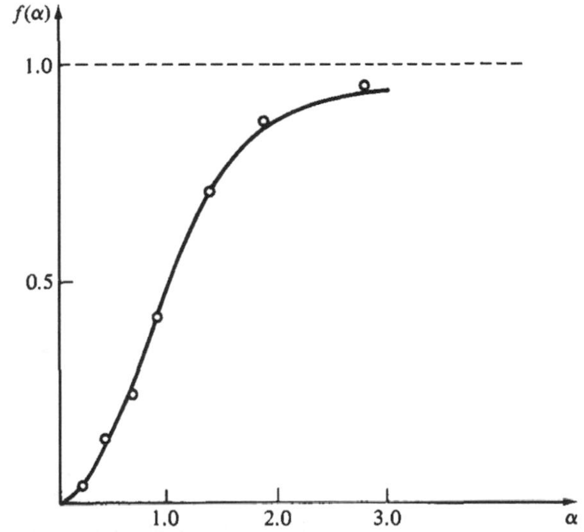

FIGURE 7.6. Saturation curve for ATCase in the absence of modifiers. The solid curve is obtained from Equation 3.8 with $n = 8$ and $U = 0.55$ and the circles are experimental values taken from Gerhart and Pardee (1963).

presence of substrate only [Gerhart and Pardee (1963)]. The value of U that best fits the data is 0.55, which is surprisingly the same value obtained for hemoglobin.

We consider now a modification of the model 3.1 to take occupation of regulatory sites into account.

In the presence of CTP, as we have noted, the initial reaction rate is inhibited as shown in Figure 7.7; i.e., the saturation curve moves off to the right in the presence of CTP, making the initial reaction rate smaller for a fixed substrate concentration in the presence of CTP.

The precise number m of regularity sites is unfortunately not known ($4 \leq m \leq 8$) and the arrangement of the regulatory subunits with respect to the catalytic subunits is also not known. We consider here only one possible

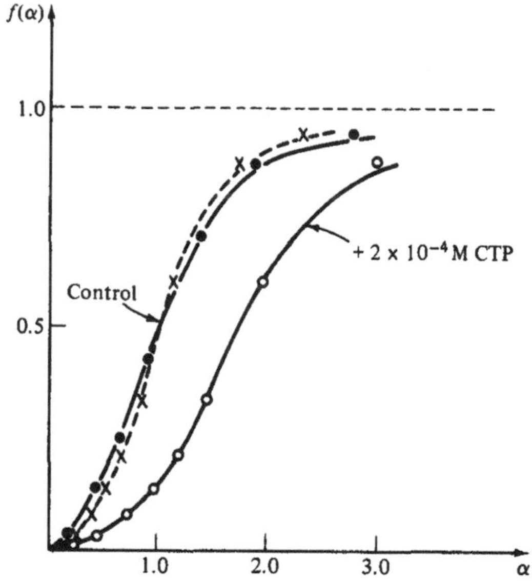

FIGURE 7.7. Saturation curves for ATCase. Black circles are experimental values for control (no CTP) and white circles are experimental values for ATCase in the presence of 2×10^{-4} M CTP. The solid curve for control is obtained from Equation 3.8 with $n = 8$ and $\exp(-4U) = 0.11$; the solid CTP curve is obtained from Equation 3.8 with α and U replaced by $\tilde{\alpha}$ and $\bar{U}$ (Equations 5.9 and 5.10) and $n = 8$ and $\exp(-4\bar{U}) = 0.10$. The dashed curve is the CTP curve rescaled so that half-saturation occurs at concentration unity. The crosses are the rescaled experimental values.

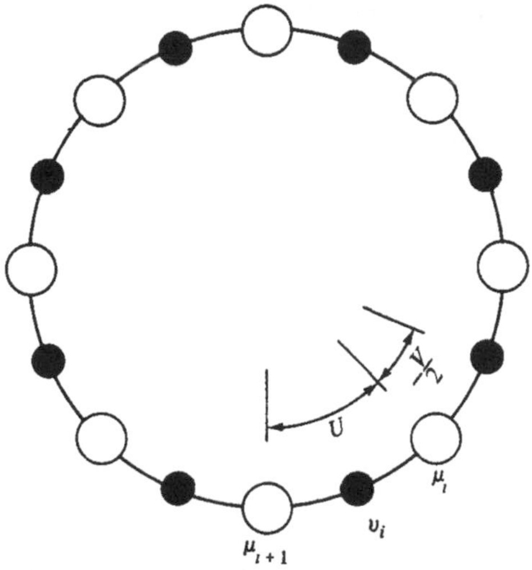

FIGURE 7.8. ATCase model. The white subunits bind substrate and the black subunits bind modifiers (CTP). Neighboring white subunits and neighboring black and white subunits only are assumed to interact.

model, based on the original experiments, which assumes eight catalytic and eight regulatory subunits, as shown in Figure 7.8. As discussed above, assuming four, six, or eight of each will not make any appreciable difference to the saturation curves (the error in any case is probably significantly less than the experimental error).

To take account of CTP in the model shown in Figure 7.8 we introduce a set of additional parameters v_i, $i = 1, 2, \ldots, n$ with values $+1$ if the ith regulatory site [between the ith and $(i + 1)$th catalytic sites] is occupied by CTP and -1 if the ith regulatory site is unoccupied by CTP. The parameters μ_i, $i = 1, 2, \ldots, n$, as before, specify the states of the catalytic sites and a configuration of the molecule is specified in the presence of CTP by

$$\{\mu; v\} = (\mu_1, \mu_2, \ldots, \mu_n; v_1, v_2, \ldots, v_n). \tag{5.2}$$

We take the probability distribution for such a configuration to be

$$P\{\mu; v\} = Z^{-1} \prod_{i=1}^{n} (e^{J\mu_i} e^{U\mu_i\mu_{i+1}})[e^{I(1+v_i)} e^{-V/4(1+v_i)(\mu_i+\mu_{i+1})}], \tag{5.3}$$

where for normalization

$$Z = \sum_{\{\mu; v\}} \prod_{i=1}^{n} (e^{J\mu_i} e^{U\mu_i\mu_{i+1}})[e^{I(1+v_i)} e^{-V/4(1+v_i)(\mu_i+\mu_{i+1})}]. \tag{5.4}$$

When all $\nu_i = -1$, i.e., all regulatory sites are unoccupied, Equation 5.3 reduces, as it should, to the previous model Equation 3.1. The factor $\exp[I(\nu_i + 1)]$ in Equation 5.3 represents independent binding on the regulatory sites and the last factor, $\exp[-(V/4)(1 + \nu_i)(\mu_i + \mu_{i+1})]$ with $V > 0$, represents repulsive interactions between nearest-neighbor regulatory and catalytic sites. We have not included any interactions between neighboring regulatory sites since it is found experimentally that ATCase, in the presence of CTP only, fits the classical Michaelis–Henri equation; i.e., there are no apparent interactions among regulatory sites.

Note that the "interaction term" in Equation 5.3,

$$U\mu_i\mu_{i+1} - \frac{V}{4}(1 + \nu_i)(\mu_i + \mu_{i+1}) = U - V \tag{5.5}$$

when $\mu_i = \mu_{i+1} = \nu_i = +1$, i.e., when the ith regulatory site and its two neighboring catalytic sites *are all occupied*. This means that when $U = V$, which, in fact, turns out to be the choice that best fits the data, the repulsive interaction between *occupied* neighboring regulatory and catalytic sites, completely annihilates the attractive interaction between *occupied* neighboring catalytic sites. In other words, $U = V$ for the present model corresponds to complete inhibition.

The evaluation of Z (Equation 5.4), which in magnetic language is the partition function for a "decorated Ising model" (or an Ising magnet with spin impurities), is accomplished in much the same way as before with the aid of the transfer-matrix method. Thus, if we first sum over the ν variables in Equation 5.4 we obtain

$$Z = \sum_{\{\mu\}} \prod_{i=1}^{n} L'(\mu_i, \mu_{i+1}), \tag{5.6}$$

where

$$L'(\mu_i, \mu_{i+1}) = e^{J\mu_i/2} e^{U\mu_i\mu_{i+1}} [1 + e^{2I} e^{-V/2(\mu_i + \mu_{i+1})}] e^{J\mu_{i+1}/2},$$

or, in matrix form,

$$\mathbf{L}' = \begin{matrix} & \mu_{i+1} = +1 & \mu_{i+1} = -1 \\ \mu_i = +1 & e^{U+J}(1 + \beta e^{-V}) & e^{-U}(1 + \beta) \\ \mu_i = -1 & e^{-U}(1 + \beta) & e^{U-J}(1 + \beta e^{V}) \end{matrix}$$

$$= (1 + \beta)e^{\bar{U} - U} \begin{pmatrix} e^{\bar{U}+J} & e^{-\bar{U}} \\ e^{-\bar{U}} & e^{\bar{U}-J} \end{pmatrix}, \tag{5.7}$$

where

$$\beta = e^{2I} \tag{5.8}$$

is a measure of the CTP concentration, $\alpha = e^{2J}$ (as before) is a measure of the substrate concentration, and $\bar{U}$ and $\bar{J}$ are defined by

$$\begin{aligned}\bar{\alpha} &= e^{2\bar{J}} \\ &= \alpha \frac{1 + \beta e^{-V}}{1 + \beta e^{V}}\end{aligned} \tag{5.9}$$

and

$$e^{4\bar{U}} = e^{4U}\left[1 + \frac{4\beta}{(1 + \beta)^2}\sinh^2\left(\frac{V}{2}\right)\right]. \tag{5.10}$$

Apart from the factor $C = (1 + \beta)\exp(\bar{U} - U)$, which is *independent of the substrate concentration* α, the matrix $\mathbf{L}'$ is identical with the matrix $\mathbf{L}$ (Equation 3.5) with J and U replaced by $\bar{J}$ and $\bar{U}$, respectively. It follows that the (two) eigenvalues of $\mathbf{L}'$ are, apart from the constant C, given by Equation 3.6 with J and U replaced by $\bar{J}$ and $\bar{U}$, respectively. The argument leading to Equation 3.7 for the average number of occupied catalytic sites can now be repeated. Since C is independent of J it cancels in Equation 3.7 and we find that the fraction of occupied catalytic sites $f(\alpha)$ is given by Equation 3.8 with α and U replaced by $\bar{\alpha}$ and $\bar{U}$ defined by Equations 5.9 and 5.10, respectively.

Half-saturation occurs when $\bar{\alpha} = 1$ [since $f(1) = \frac{1}{2}$], i.e., from Equation 5.9, when

$$\alpha = \alpha_{1/2} = \frac{1 + \beta e^{V}}{1 + \beta e^{-V}}. \tag{5.11}$$

Since V is assumed to be positive, $\alpha_{1/2}$ is an increasing function of CTP concentration β with the limiting value

$$\lim_{\beta \to \infty} \alpha_{1/2} = e^{2V}. \tag{5.12}$$

The increase of $\alpha_{1/2}$ with increasing CTP concentration is observed experimentally [Figure 7.7, Gerhart and Pardee (1963)] and so is the limiting behavior described by Equation 5.12. Notice that in our model the limiting value of $\alpha_{1/2}$ is directly related to the interaction parameter V between regulatory and catalytic sites.

It is clear from Equation 5.10 that $\bar{U} \geq U$; i.e., in the presence of CTP the "effective interaction" between catalytic sites is increased. Notice, however, from Equation 5.10, that $\bar{U}$ as a function of β achieves a maximum value $\bar{U}_M$ given by

$$e^{4\bar{U}_M} = e^{4U}\left[1 + \sinh^2\left(\frac{V}{2}\right)\right] \tag{5.13}$$

when $\beta = 1$, and that $\lim_{\beta\to\infty} \bar{U} = U$. In other words, according to the model, cooperativity should at first increase with increasing CTP concentration until it reaches a maximum at a certain concentration $\beta_M(=1$ in our units), and for $\beta > \beta_M$ cooperativity should decrease until in the limit $\beta \to \infty$, $\bar{U} = U$, which is also the case when $\beta = 0$. This effect should be clearly visible if the saturation curves for different CTP concentrations are plotted on a scale so that half-saturation always occurs at concentration unity (see Figure 7.7). Unfortunately, the data necessary to test this prediction do not seem to exist. The only published data for ATCase of the desired type [Gerhart and Pardee (1963)] give two saturation curves (shown in Figure 7.7), one for ATCase in the absence of modifiers ("control") and one in the presence of 2×10^{-4} M CTP. These two curves are sufficient, however, to determine all the parameters in the model (U, V, and the scale of concentration β).

With the control $\alpha_{1/2} = 1$ we have found that the choice $U = 0.55$ best fits the control data. In the presence of 2×10^{-4} M CTP, $\alpha_{1/2} \cong 1.7$ (from Figure 7.7), so from Equation 5.11 we obtain

$$\frac{1 + \beta e^{V}}{1 + \beta e^{-V}} \cong 1.7. \tag{5.14}$$

Furthermore, if we rescale the CTP curve as described above (shown as a dashed curve in Figure 7.7) and fit the resulting curve to Equation 3.8 we obtain $\exp(-4\bar{U}) \cong 0.10$. Since $\exp(-4U) \cong 0.11$ we find from Equation 5.10 that

$$\frac{4\beta}{(1 + \beta)^2} \sinh^2\left(\frac{V}{2}\right) \cong 0.10. \tag{5.15}$$

Solving Equations 5.14 and 5.15 for β and V we obtain

$$\beta \cong 0.93 \tag{5.16}$$

and

$$\exp(2V) \cong 3.0. \tag{5.17}$$

The values of β and V are unfortunately very sensitive to the values of $\bar{U}$ and U, so Equations 5.16 and 5.17 could conceivably be in error by as much as 20 percent or so (if $\bar{U}$ and U are in error by a few percent).

Nevertheless, from Equations 5.17 and 5.12 we conclude that the ratio of $\alpha_{1/2}$ in the limit of infinite CTP concentration, to $\alpha_{1/2}$ for no CTP, should be approximately 3. Also, from Equation 5.16 and the above discussion, we conclude that maximum cooperativity should be achieved at a CTP concentration of about 2×10^{-4} M. Unfortunately, the data to test either of these predictions do not seem to exist.

One interesting result, which emerges from Equation 5.17, is that $V \approx U \approx 0.55$, which means, as remarked above, that in the model, at least, we have complete inhibition.

The model described above can be generalized in any number of ways, for example, to take activators, which shift the curve to the left, as well as inhibitors into account on ATCase. Models can also be formulated for allosteric enzymes that do not have distinct subunits for substrate and modifiers (e.g., DPN–isocitrate dehydrogenase). Some discussion of these cases can be found in Thompson (1968b) (see Problem 3).

7-6 Ising Models for DNA

Deoxyribonucleic acid (DNA for short) is the blueprint for life, the carrier of the genetic code. Naturally occurring DNA is a large molecule, typically 30,000 Å long and 20 Å thick, composed of some 1 million atoms. It contains four essential components, called bases: adenine (A), thymine (T), cytosine (C), and guanine (G); and sections of the molecule from a given organism, consisting of sequences of A's, T's, C's, and G's, are claimed to determine the set of proteins that the organism can make.

According to the Watson–Crick model [Watson (B1968)] DNA is composed of two strands wrapped around one another to form a double helix, as shown in Figure 7.9. There are sequences of the four bases on each strand which are connected to one another by (relatively weak) hydrogen bonds, also shown in Figure 7.9. An important feature of the model is that A can only be coupled to T, and C can only be coupled to G, so AT and CG are the only possible base pairs. In the model this is necessary to obtain a compact structure. The occurrence of equal amounts of A and T and of C and G was,

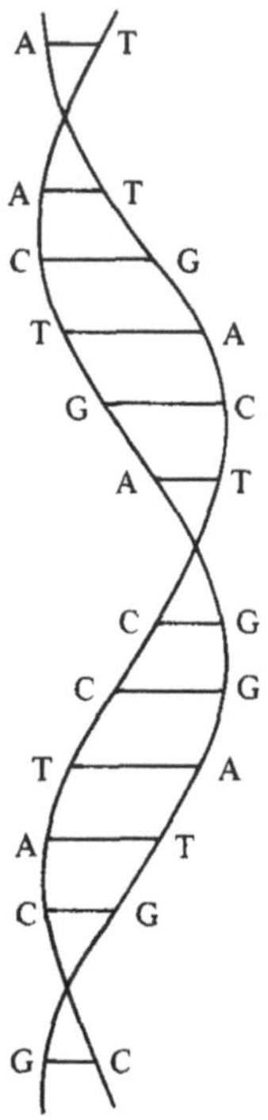

FIGURE 7.9. Watson–Crick model for DNA.

however, discovered by Chargaff (1950) before the invention of the Watson–Crick model and is known as Chargaff's rule.

Because of Chargaff's rule and the two-stranded structure, the sequence of bases on one strand completely determines the complementary sequence of bases on the other strand. This has very important genetic implications. Replication of DNA, for example, can be achieved in the model by unwinding the molecule and making two new complementary strands from free bases. The sequence of bases on one strand (or its complementary sequence) is also claimed to be the blueprint for the production of proteins.

Since the model is so simple and explains almost everything that needs to be explained, it has been almost universally accepted. Its inventors, Crick and Watson, and also Wilkins, who provided the X-ray pictures that suggested a double-helix structure, were justly rewarded with the Nobel prize for medicine in 1962—the same year, in fact, that Perutz and Kendrew were awarded the chemistry prize for the structure of hemoglobin and myoglobin. Some very recent work, however, suggests that the structure of DNA may not be quite as simple as that proposed by Watson and Crick. Wu (1969), in particular, has made a very careful reevaluation of the X-ray pictures (at high humidity) and finds that the data fit a four-strand model much better than a two-strand model. The final word on the subject, however, will have to wait until more detailed X-ray pictures become available.

The considerable amount of theoretical work that has been done on the Watson–Crick model has been mainly concerned with the denaturation process, i.e., the breaking of the hydrogen bonds connecting the two strands under treatment by heat or other things, such as pH. As the temperature is increased, for example, bonds break until finally one is left with two separate strands or coils. The denaturation process, or helix-coil transition, produces melting curves (the fraction of broken bonds as a function of temperature) with the characteristic sigmoid shape shown in Figure 7.10, reflecting some degree of cooperativity or interaction among the hydrogen bonds connecting the two strands. The temperature T_1 where one half of the bonds are broken is called the melting point, and around this point the bonds break rather easily.

The first theoretical treatment of the denaturation process was given by

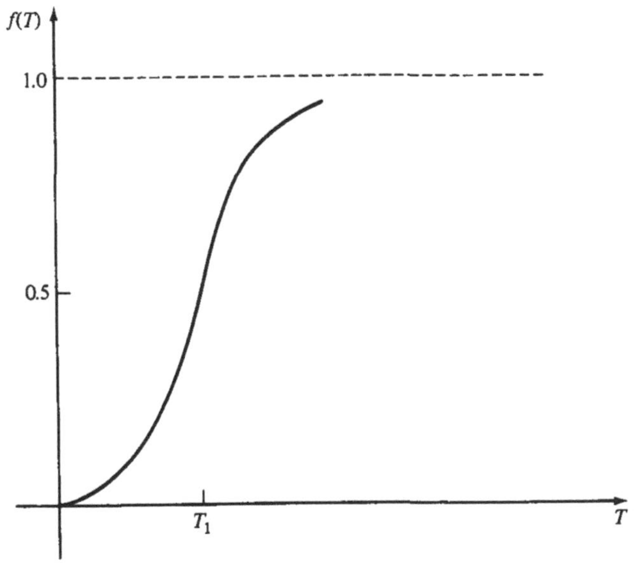

FIGURE 7.10. Fraction of broken bonds $f(T)$ as a function of temperature T. One half of the bonds are broken at the melting point T_1.

Zimm and Bragg (1959), who assumed that all bonds are identical. This is surely not the case in naturally occurring DNA, since AT pairs are connected by two hydrogen bonds, while CG pairs are connected by three hydrogen bonds. Synthetic one-component DNA's consisting of all AT or all CG pairs can be made, however, and the Zimm–Bragg model should be applicable to such molecules. We begin, therefore, with a discussion of one-component DNA. The presentation given by Montroll and Goel (1966) and Goel and Montroll (1968) will be followed throughout, since they use Ising-model language exclusively. Various periodic synthetic two-component (i.e., AT and CG pairs) DNA molecules are also discussed by Montroll and Goel, and we shall have a little to say about these at the end of this section. The interested reader is referred to Montroll and Goel (1966) and Goel and Montroll (1968) for lists of references and relations between previous work and the models to be discussed here.

For all intents and purposes we can consider the Watson–Crick model as the ladder shown in Figure 7.11, the sides of the ladder corresponding to the two strands and the rungs of the ladder to the complexes joining the base pairs. By complexes we mean the two hydrogen bonds connecting an AT

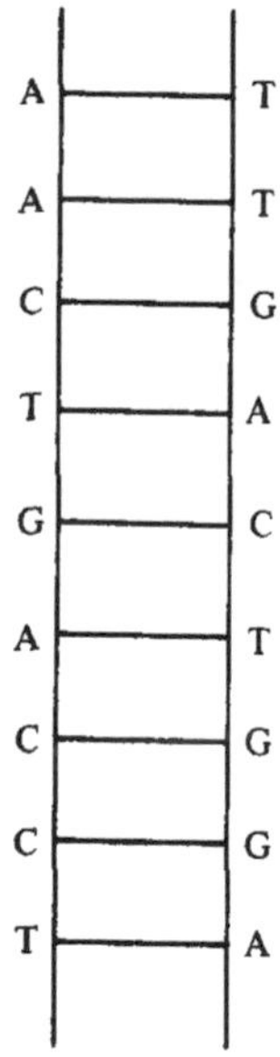

FIGURE 7.11. Ladder version of the Watson–Crick model (Figure 7.9).

pair and the three hydrogen bonds connecting a CG pair. We number the complexes by an index $i = 1, 2, \ldots, n$ and assume for simplicity that a complex is either broken or intact. We then assign a parameter μ_i to the ith complex with values

$$\mu_i = \begin{cases} +1 & \text{if the } i\text{th complex is broken} \\ -1 & \text{if the } i\text{th complex is intact.} \end{cases} \tag{6.1}$$

In a given configuration of broken and intact complexes $\{\mu\} = (\mu_1, \ldots, \mu_n)$, the number of broken complexes is given by

$$N\{\mu\} = \sum_{i=1}^{n} \tfrac{1}{2}(1 + \mu_i), \tag{6.2}$$

and if $P\{\mu\}$ is the probability distribution for the configuration $\{\mu\}$, the average number of broken complexes N is given by

$$N = \sum_{\{\mu\}} N\{\mu\}P\{\mu\}, \tag{6.3}$$

where the sum is over all (2^n) possible configurations $\mu_1 = \pm 1$, $\mu_2 = \pm 1, \ldots,$ $\mu_n = \pm 1$ of the molecule.

The formulation given so far is completely general and is equivalent to the formulation given in previous sections for hemoglobin and (allosteric) enzymes. The problem, as before, is to specify the probability distribution $P\{\mu\}$.

Consider first the random one-component case, i.e., with probability distribution $P\{\mu\}$ given by Equations 2.7 and 2.8,

$$P\{\mu\} = \prod_{i=1}^{n} p(\mu_i), \tag{6.4}$$

$$p(\mu_i) = (2 \cosh J)^{-1} \exp(J\mu_i), \tag{6.5}$$

where now $p(+1)$ is the probability that a complex is broken and $p(-1)$ is the probability that a complex is intact. In this case the fraction of broken complexes f is given by Equation 2.12, i.e.,

$$f = \frac{N}{n} \tag{6.6}$$

$$= \tfrac{1}{2}(1 + \tanh J).$$

To determine the temperature dependence of f we must determine the temperature dependence of J. The obvious statistical-mechanical choice, from Equation 6.5, is $J = E/kT$, where E is the energy difference between bonded and unbonded states. In the limit of infinite temperature ($J = 0$), however, Equation 6.6 then gives $f = \frac{1}{2}$, which is certainly incorrect since we know from experiment that all bonds are broken at sufficiently high temperatures. E, defined above, must then be temperature dependent if we wish to imitate Figure 7.10 with Equation 6.6. We therefore make the empirical choice

$$J = a(T - T_1). \tag{6.7}$$

$f = f(T)$ is then one half at the melting point T_1 and Equation 6.6 gives the curve shown in Figure 7.12.

The slope of the curve at $T = T_1$ from Equations 6.6 and 6.7 is obviously $a/2$, so the melting curve can be made more "cooperative" or less "cooperative" by increasing or decreasing a.

For pure synthetic AT-DNA, Equations 6.6 and 6.7 give a reasonably good fit to experiment with the choice $a = 1$ and $T_1 = 65°C$.

The random model, however, is a little unrealistic and does not in general give a good fit to experiment. A more reasonable choice for the probability

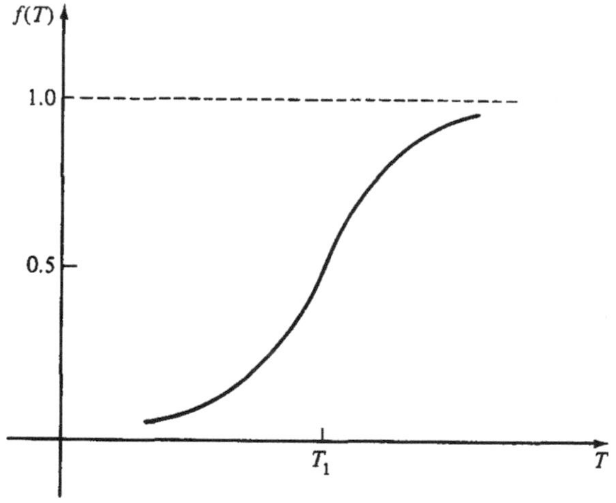

FIGURE 7.12. Melting curve obtained from the random model Equations 6.6 and 6.7.

distribution $P\{\mu\}$, which takes nearest-neighbor complex interactions into account, is (Equation 3.1)

$$P\{\mu\} = Z^{-1} \prod_{i=1}^{n} e^{J\mu_i} e^{U\mu_i\mu_{i+1}}, \tag{6.8}$$

where

$$Z = \sum_{\{\mu\}} \prod_{i=1}^{n} e^{J\mu_i} e^{U\mu_i\mu_{i+1}} \tag{6.9}$$

is again the Ising-model partition function, and for simplicity we have taken $\mu_{n+1} = \mu_1$, i.e., "circular DNA" (since n is large, boundary, or free end, terms are unimportant).

The fraction of broken complexes from Equations 3.6 and 3.7 is given by

$$f = \frac{N}{n} = \frac{1}{2}\left[1 + (\lambda_1^n + \lambda_2^n)^{-1}\left(\lambda_1^{n-1}\frac{\partial\lambda_1}{\partial J} + \lambda_2^{n-1}\frac{\partial\lambda_2}{\partial J}\right)\right], \tag{6.10}$$

where

$$\left.\begin{matrix}\lambda_1\\ \lambda_2\end{matrix}\right\} = e^{U}\cosh J \pm (e^{-2U} + e^{2U}\sinh^2 J)^{1/2}. \tag{6.11}$$

In the limit $n \to \infty$ only the largest eigenvalue λ_1 contributes to Equation 6.10, giving

$$f = \lim_{n\to\infty}\frac{N}{n} = \frac{1}{2}\left[1 + \frac{\sinh J}{(e^{-4U} + \sinh^2 J)^{1/2}}\right]. \tag{6.12}$$

In the limit $U = 0$ (i.e., no interactions) Equation 6.12 reduces to the random model equation (6.6) and in the limit $U \to \infty$, the "zipper limit," f becomes a step function; i.e., $f = 1$ when $J > 0$ and $f = 0$ when $J < 0$, as shown in Figure 7.13 for the choice $J = a(T - T_1)$ (Equation 6.7).

In general the model gives a reasonably good fit to experiment, but it has one serious deficiency—it does not take long sequences of broken bonds into account in a realistic way. As bonds break, rings or loops are formed by the separated strands. Such rings can exist in a number of possible geometrical configurations, giving rise, in statistical-mechanical language, to a ring entropy. Also it is reasonable to suppose that bonds will break more easily in the neighborhood of a loop. Neither of these effects is taken into account in Equation 6.8 for the probability distribution $P\{\mu\}$.

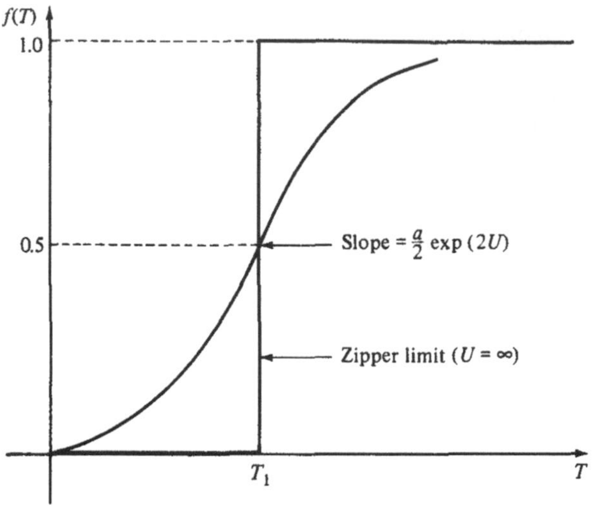

FIGURE 7.13. Melting curves obtained from Equation 6.12.

Various models have been proposed to take account of the ring entropy [see Goel and Montroll (1968) and references therein], but none is entirely satisfactory. It should be noted, however, that the above model is probably a reasonably good model for temperatures below the melting point, where large rings or loops are not yet formed.

The Ising model can in principle be modified to take loops and different complexes (AT and CG) into account, but the mathematical manipulation required then becomes extremely unwieldy. One interesting question, which has not been satisfactorily answered, is: Can one deduce any (statistical) information about the distribution of base pairs in naturally occurring DNA from the melting curves? In an attempt to answer this question Goel and Montroll considered various synthetic DNA's with known periodic distribution of AT's and CG's. One obtains a variety of melting curves for such molecules (some examples are shown in Figure 7.14) and the hope is that by considering simple models for such molecules, model parameters will be found to fit more complicated DNA forms.

One interesting fact that appears to emerge from these studies is that the coupling constant U is temperature independent and is essentially the same for nearest-neighbor AT-AT, AT-GC, or GC-GC complexes. Assuming that U is, in fact, independent of temperature and complex, a probability dis-

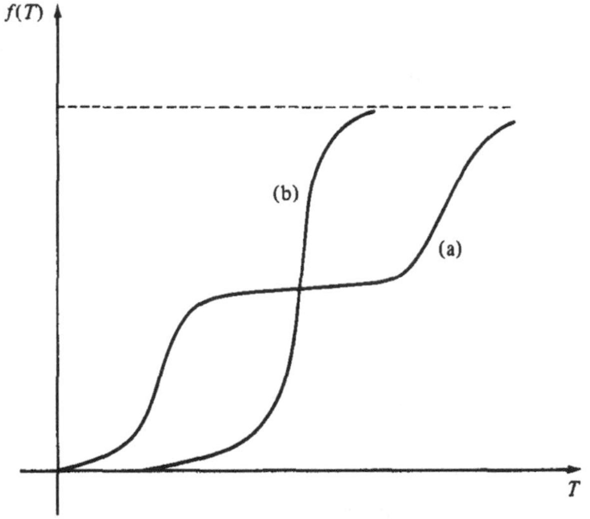

FIGURE 7.14. Melting curves for DNA:
(a) AAA ··· AGG ··· G / TTT ··· TCC ··· C
(b) AGAG ··· / TCTC ···

tribution $P\{\mu\}$ appropriate for the sequence of AT's and CG's shown in Figure 7.15 is

$$P\{\mu\} = Z^{-1}\left(\prod_{i=1}^{n} e^{U\mu_i\mu_{i+1}}\right)e^{J_1(\mu_1+\cdots+\mu_l)}e^{J_2(\mu_{l+1}+\cdots+\mu_{l+m})}$$
$$\times\, e^{J_1(\mu_{l+m+1}+\cdots+\mu_{l+m+s})}\cdots, \tag{6.13}$$

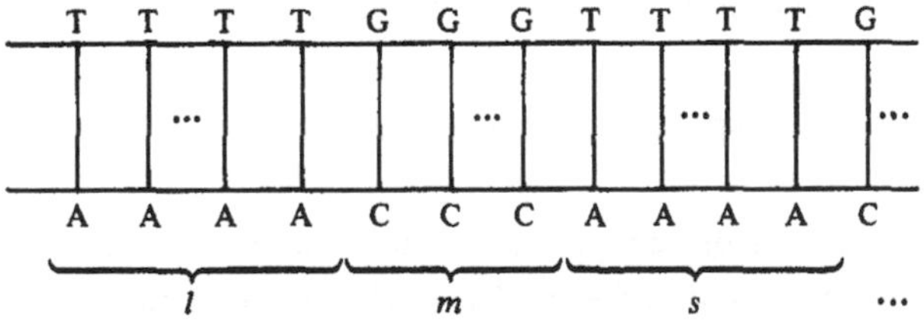

FIGURE 7.15. Periodic two-component DNA.

where J_1 refers to AT complexes and J_2 to CG complexes. The evaluation of the fraction of broken bonds for such a distribution can be carried out with the aid of transfer-matrix methods (see Problem 4 for some special cases) and, if one wishes, loops can be included as well.

PROBLEMS

1. Derive Equation 4.6 for the fraction of occupied sites in the Monod–Wyman–Changeux model.

2. Construct a probability distribution for the following generalized Monod–Wyman–Changeux model:

(a) The whole molecule can exist in either an R state or a T state;

(b) Substrate (S) binds exclusively to molecules in the R state;

(c) Inhibitor (I) binds exclusively to molecules in the T state;

(d) Activator (A) binds exclusively to molecules in the R state. Assuming that there are n binding sites for S, I, and A, show that the fraction of sites occupied by substrate is given by

$$\frac{\alpha(1+\alpha)^{n-1}(1+\gamma)^n}{(1+\alpha)^n(1+\gamma)^n + L(1+\beta)^n},$$

where L is the allosteric constant (Equation 4.2) and α, β, and γ are the concentrations of substrate, inhibitor, and activator, respectively.

3. Consider a molecule composed of n subunits with one binding site for substrate and one binding site for inhibitor on each subunit. Assuming that only nearest-neighbor substrate sites interact and that occupation of a subunit by inhibitor affects only the substrate occupation of that subunit, show that the probability distribution given by (see Equation 5.3)

$$P\{\mu;\nu\} = Z^{-1}\prod_{i=1}^{n}\exp\left[J\mu_i + U\mu_i\mu_{i+1} + I(1+\nu_i) - \frac{V}{2}(1+\nu_i)\mu_i\right]$$

leads to the fraction of occupied substrate sites $f(\bar{\alpha})$ given by Equation 3.8 with $\bar{\alpha}$ defined by Equation 5.9 (and U unchanged). (This is a possible model for DPN–isocitrate dehydrogenase [Thompson (1968b)].)

4. The probability distribution for periodic AT, CG, AT, CG, ... DNA, excluding loop effects and assuming only nearest-neighbor complex interactions, is given by (Equation 6.13)

$$P\{\mu\} = Z^{-1}\left(\prod_{i=1}^{2n} e^{U\mu_i\mu_{i+1}}\right)e^{J_1(\mu_1+\mu_3+\cdots+\mu_{2n-1})}\,e^{J_2(\mu_2+\mu_4+\cdots+\mu_{2n})},$$

where J_1 refers to AT complexes and J_2 to CG complexes. Use the transfer-matrix method to calculate the average number of broken bonds, assuming that $J_1 = a(T - T_1)$ and $J_2 = a(T - T_2)$.

Do the same for AT-AT-AT $\cdots$ AT-CG-CG $\cdots$ CG DNA, assuming only one interaction parameter U for nearest-neighbor complexes.

Appendixes

APPENDIX A

Measure-theoretic Statement of Liouville's Theorem

We shall follow the discussion given by Khinchin (B1949).

Let A be any (Lebesque) measurable set of Γ points and let A_t denote the time-development set obtained from A according to Hamilton's equations after time t.

Define the *measure* of A by

$$V(A) = \int_A d\Gamma = \int_A dq_1\, dq_2 \cdots dq_n\, dp_1\, dp_2 \cdots dp_n, \tag{1}$$

where n denotes the number of degrees of freedom. The rest of the notation is the same as in Chapter 1 (Section 1-7 especially).

We now prove

LIOUVILLE'S THEOREM:

$$V(A_t) = V(A). \tag{2}$$

Changing variables in the integral $V(A_t)$ (Equation 1) to $q_1(0), \ldots, q_n(0)$, $p_1(0), \ldots, p_n(0)$, i.e., the initial values, we have that

$$V(A_t) = \int_A J(t; q_1(0), \ldots, p_n(0))\, dq_1(0) \cdots dp_n(0), \tag{3}$$

where

$$J(t; q_1(0), \ldots, p_n(0)) = \frac{\partial(q_1, \ldots, q_n, p_1, \ldots, p_n)}{\partial(q_1(0), \ldots, q_n(0), p_1(0), \ldots, p_n(0))} \tag{4}$$

is the Jacobian of the transformation.

For simplicity we now use the notation

$$\left.\begin{array}{ll} q_i = x_i, & p_i = x_{i+n} \\ q_i(0) = y_i, & p_i(0) = y_{i+n} \end{array}\right\} 1 \leq i \leq n. \tag{5}$$

Now, from the rule for differentiating determinants,

$$\frac{\partial J}{\partial t} = \sum_{i=1}^{2n} J_i, \tag{6}$$

where

$$J_i = \frac{\partial(x_1, \ldots, x_{i-1}, \dot{x}_i, x_{i+1}, \ldots, x_{2n})}{\partial(y_1, \ldots, y_{i-1}, y_i, y_{i+1}, \ldots, y_{2n})} \qquad \left(\dot{x}_i = \frac{dx_i}{dt}\right). \tag{7}$$

The (i, k) element in the determinant J_i is

$$\frac{\partial \dot{x}_i}{\partial y_k} = \sum_{j=1}^{2n} \frac{\partial \dot{x}_i}{\partial x_j}\frac{\partial x_j}{\partial y_k};$$

hence

$$\begin{aligned} J_i &= \sum_{j=1}^{2n} \frac{\partial \dot{x}_i}{\partial x_j} \frac{\partial(x_1, \ldots, x_{i-1}, x_j, x_{i+1}, \ldots, x_{2n})}{\partial(y_1, \ldots, y_{i-1}, y_i, y_{i+1}, \ldots, y_{2n})} \\ &= J\frac{\partial \dot{x}_i}{\partial x_i} \end{aligned} \tag{8}$$

since the Jacobian in the summation on j is obviously zero unless $j = i$. From Equation 6, therefore,

$$\begin{aligned} \frac{\partial J}{\partial t} &= J\sum_{i=1}^{2n} \frac{\partial \dot{x}_i}{\partial x_i} \\ &= J \operatorname{Div} \mathbf{v} \\ &= 0, \end{aligned} \tag{9}$$

where in the second step we have used the definitions 5 and 7.11 and 7.13 of Chapter 1, and the final result follows from 7.14 of Chapter 1 or from direct application of Hamilton's equations. This completes the proof or the theorem.

Consider now the energy shell

$$E \leq H(p, q) \leq E + dE. \tag{10}$$

If A is now a subset of the *energy surface* $H(p, q) = E$, dn is the normal distance between the two energy surfaces $H(p, q) = E$ and $H(p, q) = E + dE$ and $d\sigma$ is an element of the energy surface, then Liouville's theorem states that

$$\int_{A_t} dn\, d\sigma = \int_A dn\, d\sigma. \tag{11}$$

In the notation Equation 5,

$$dx_i = dn \frac{\partial E}{\partial x_i} \|\text{grad } H\|^{-1}, \tag{12}$$

where

$$\|\text{grad } H\|^2 = \sum_{i=1}^{2n} \left(\frac{\partial H}{\partial x_i}\right)^2. \tag{13}$$

It follows that

$$dE = \sum_{i=1}^{2n} \frac{\partial E}{\partial x_i} dx_i = dn \|\text{grad } H\| \tag{14}$$

and hence from Equation 11 that

$$\mu(A_t) = \mu(A), \tag{15}$$

where $\mu(A)$ is the measure of a subset A of the whole energy surface Ω, defined by

$$\mu(A) = \frac{\int_A d\sigma \|\text{grad } H\|^{-1}}{\int_\Omega d\sigma \|\text{grad } H\|^{-1}}. \tag{16}$$

Equation 16 is precisely the definition 7.4 of Chapter 1 and Equation 15 is the form of Liouville's theorem needed to prove Poincaré's theorem (Section 1-7).

APPENDIX B

Ergodic Theory and the Microcanonical Ensemble

The main difficulty in deducing equilibrium properties of a system from the laws governing the individual particles lies in the fact that there are two distinct levels of description. Equilibrium is observed only on the macroscopic level and is described by relatively few macroscopic variables. On the microscopic level, however, one describes the motion of all the particles. It is because of this difference in description that one hopes for a reconciliation between the reversibility on the microscopic level and the apparent irreversibility on the macroscopic level.

First, we must say precisely what is meant by a macroscopic variable. Unfortunately, this notion is somewhat arbitrary and depends on how detailed a description one wants of the system. Gibbs suggests that a finite number of macroscopic variables

$$y_i = f_i(\mathbf{r}_1, \ldots, \mathbf{r}_N; \mathbf{p}_1, \ldots, \mathbf{p}_N), \qquad i = 1, 2, \ldots, M,$$

used to describe the macroscopic state of a system, should satisfy the following conditions.

1. Each macroscopic state defined by a set of y_i values corresponds to a region of Γ space, i.e., a portion of the energy surface $H(p, q) = E$.

2. For large N there is one set of values $y_1, \ldots, y_M$ which corresponds to a region that is overwhelmingly the largest.

Requirement 1 corresponds to a "coarse graining" of the energy surface and, as stressed by Ehrenfest and Ehrenfest (B1911), is essential for any macroscopic description. The amount of coarse graining clearly depends on the level of description demanded by the observer. Requirement 2, on the other hand, imposes restrictions on the type of description by requiring that N be large.

Granted conditions 1 and 2 on the macroscopic description, we now argue as follows. If every part of the energy surface is accessible to phase points, and if the time $t(A)$ a phase point spends in a region A is (roughly speaking) proportional to the area of A, we can identify the region that is overwhelmingly the largest as the macropscopic equilibrium state, since it follows that

(a) If the system is not in the equilibrium state, it will almost always go into this state.

(b) If the system is in equilibrium, it will almost always stay in equilibrium, although in view of the existence of Poincaré cycles, i.e., the quasi-periodic character of the motion in Γ space, there will be fluctuations away from the equilibrium state.

The situation is shown schematically in Figure B.1.

These ideas are due to Gibbs (similar ideas, although less precise and general, were put forward by Boltzmann) and can be made more precise by introducing probabilistic notions and appealing to the ergodic theorems.

First, we introduce the notion of probability by assuming that at time $t = 0$ the system is in a definite nonequilibrium macroscopic state specified by the y_i variables, and is characterized by a probability distribution $D(p, q, t = 0)$ which is nonzero only in the given region. $D(p, q, t = 0)$ is the phase-point

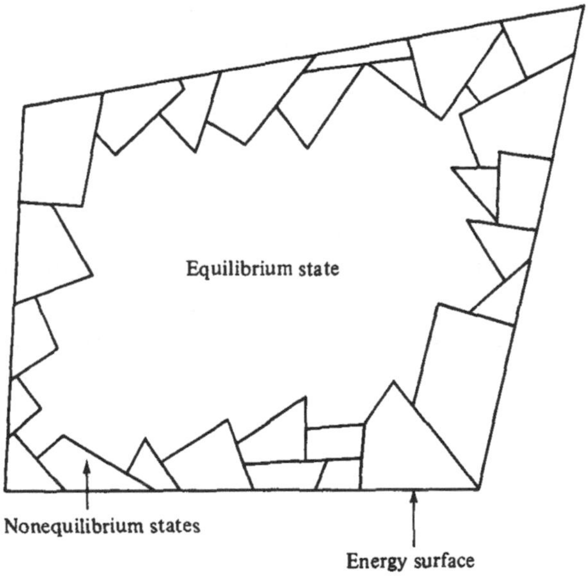

FIGURE B.1. Illustration of the partitioning of the energy surface into nonequilibrium states and an equilibrium state.

density $\rho(p, q, t = 0)$ defined previously (Equation 2.1) normalized to unity so that $\int D(p, q, t = 0)\, d\Gamma = 1$. $D(p, q, t = 0)$ is, in principle, arbitrary (since we would like equilibrium to be approached independently of the initial distribution) except that it should be a reasonably smooth function of (p, q), for example a constant over the initial region. This is the only probabilistic assumption. The time development $D(p, q, t)$ is determined from $D(p, q, t = 0)$ by Liouville's theorem, and one would like $D(p, q, t)$ to approach the equilibrium distribution as t increases.

Gibbs' description of how this might come about is as follows: From Liouville's theorem the volume of the region where $D(p, q, t)$ is nonzero remains the same, although its shape will change drastically. As time develops the initial region will be drawn out into a thin ribbon winding through the energy shell $E \leq H(p, q) \leq E + \Delta E$, until in a coarse-grained sense the distribution becomes uniform over the shell. This is Gibbs' *microcanonical distribution*. The degree of uniformity and the time it takes to achieve uniformity clearly depends on how coarse grained a description one wants. The more coarse grained the description, the shorter the time, clearly, it will take to achieve the required uniform distribution.

Although this argument is extremely heuristic it can be made mathematically more precise by appealing to the ergodic theorems. It should be stressed, however, that the basic point of the argument—the existence of a macroscopic description to begin with—is not at all solved by ergodic theory. Ergodic theorems are necessary for the validity of Gibbs' argument, but they do not answer the basic question: How is a macroscopic description possible? This is the question considered by the physicist to lie at the heart of the matter, and some physicists go as far as to say that ergodic theory is really unnecessary. Nevertheless, we conclude with a brief description of the ergodic theorems, due mainly to Birkoff. They are as follows:

1. For a bounded mechanical motion and for any phase function $y = f(p, q)$ which is integrable over the energy surface (e.g., Gibbs' macroscopic variables defined above) the time average

$$\bar{y} = \lim_{\tau \to \infty} \tau^{-1} \int_0^{\tau} f(P_t)\, dt$$

almost always exists and is independent of the initial point $P_0 = (p, q)_{t=0}$, which evolves into $P_t = (p, q)_t$ according to Liouville's theorem.

2. If the system is *metrically transitive* (i.e., if the energy surface cannot be partitioned into two disjoint regions such that points starting from one region will always remain in that region), the time average $\bar{y}$ is almost always equal to the ensemble average

$$\langle y \rangle = \int_{H=E} \cdots \int f(P)\sigma(H)\, d\Gamma$$

over the energy surface with respect to the uniform distribution (i.e., the microcanonical distribution)

$$\sigma(H) = \left(\|\text{grad } H\| \int_{H=E} \cdots \int \|\text{grad } H\|^{-1}\, d\Gamma\right)^{-1}$$

on the energy surface.

Taking $f(P)$ in particular to be the characteristic function for a region A, i.e., $f(P) = 1\ P \in A,\ 0\ P \notin A$, described by the probability density $D(p, q, t)$, the ergodic theorems show that for a metrically transitive system

$$\lim_{T\to\infty} \frac{t(A)}{T} = \frac{V(A)}{V},$$

where $t(A)$ is the time a given point spends in A, $V(A)$ is the volume of A, and V is the volume of the whole energy shell.

Clearly this is a precise formulation of the basic step in the argument—that all points are accessible and the time a point spends in a region is determined by its area. A necessary prerequisite then for the Gibbs microcanonical distribution is that the system be metrically transitive. Until very recently there were no nontrivial physical examples of metrically transitive systems. Sinai (1963, 1966), however, has recently proved that (three or more) hard spheres in a box form a metrically transitive system. (He in fact proves a stronger result; i.e., for those who know the jargon, hard spheres in a box form a K system.) It is interesting to note that in this example, metric transitivity exists for a *finite number* of particles. Previously it was felt that an infinite number would be required foı the existence of metric transitivity. Sinai has also very recently extended his result to hard spheres with a finite range attraction (with rather restrictive conditions, however), so more faith can now be placed in Gibbs' postulate. We stress again, however, that important as these results are, they do not solve the basic physical problem of the existence of macroscopic description.

APPENDIX C

Lebowitz–Penrose Theorem for Magnetic Systems

For simplicity we shall consider a one-dimensional chain of N spins $\mu_1, \mu_2, \ldots, \mu_N$ ($\mu_i = \pm 1$) with interaction energy

$$E\{\mu\} = - \sum_{1 \le i < j \le N} \phi(|i-j|)\mu_i \mu_j, \tag{1}$$

where $\varphi(x) = \gamma v(\gamma|x|)$ and $v(y)$ is such as to ensure the existence of the thermodynamic limit. It is convenient to write Equation 1 in the symmetrical form

$$E\{\mu\} = - \sum_{i \ne j=1}^{N} \phi(|i-j|)\left(\frac{\mu_i + \mu_j}{2}\right)^2 + \frac{1}{2} \sum_{i \ne j=1}^{N} \phi(|i-j|), \tag{2}$$

making use of the result $\mu_i^2 = 1$. Since the second term in Equation 2 is a trivial constant we shall ignore it henceforth.

To prove the lattice analogue of the Lebowitz–Penrose theorem stated in Section 4-4, we consider the partition function with a fixed number of down spins N_- ($\mu_i = -1$), defined by

$$Z(N, N_-) = \sum_{\{\mu\}}{}' \exp\left[\beta\gamma \sum_{i \ne j} v(\gamma|i-j|)\left(\frac{\mu_i + \mu_j}{2}\right)^2\right], \tag{3}$$

where the restricted sum is over configurations $\{\mu\}$ with a fixed number N_- of down spins, i.e.,

$$\sum_{i=1}^{N} \mu_i = N - 2N_-, \tag{4}$$

and for convenience we have suppressed the temperature dependence of the partition function.

In the gas interpretation (see Sections 4-5 and 5-2), N_- is considered as the number of particles and N the "volume." $\rho = N_-/N$ is then the density.

The theorem we shall prove is the following (see Equation 4.20 of Chapter 4):

THEOREM. *For the partition function $Z(N, N_-)$ defined by Equation* 3 *with* $v(x) \geq 0$,

$$\lim_{\gamma \to 0} \lim_{\substack{N, N_- \to \infty \\ \rho = N_-/N \text{ fixed}}} N^{-1} \log Z(N, N_-)$$
$$= CE\{\nu[\rho^2 + (1-\rho)^2] - \rho \log \rho - (1-\rho)\log(1-\rho)\} \quad (5)$$

where CE denotes convex envelope (see Figure C-2), $\nu = \beta\alpha$, and

$$\alpha = \int_0^\infty v(x)\, dx < \infty. \quad (6)$$

(Note that Equation 5 is the free energy per volume in the gas language.) The Curie–Weiss result (Equation 5.19 of Chapter 4) is obtained from Equation 5 as follows.

With interaction energy

$$E\{\mu\} = -\gamma \sum_{i \neq j} v(\gamma |i-j|)\left(\frac{\mu_i + \mu_j}{2}\right)^2 - H \sum_{i=1}^{N} \mu_i. \quad (7)$$

The partition function $Z(N)$ can be written as

$$\begin{aligned} Z(N) &= \sum_{\{\mu\}} \exp(-\beta E\{\mu\}) \\ &= \sum_{N_-=0}^{N} \exp[B(N - 2N_-)] Z(N, N_-), \; (B = \beta H), \end{aligned} \quad (8)$$

where $Z(N, N_-)$ is the restricted partition function defined by Equation 3. In the proof of the theorem we will see that $\log Z(N, N_-)$ is proportional to N in the limit $N \to \infty$; hence

$$Z(N) \sim \max_{0 \leq N_- \leq N} \{Z(N, N_-)\exp[NB(1 - 2N_-/N)]\}. \quad (9)$$

Moreover, the maximum in Equation 9 occurs for N_- proportional to N, so from Equation 5 and the fact that the maximum of the convex envelope of a function is simply the maximum of the function (see Figure C.2), we have that

$$\lim_{\gamma \to 0} \lim_{N \to \infty} N^{-1} \log Z(N) = \max_{0 \leq \rho \leq 1} \{\nu[\rho^2 + (1-\rho)^2] + B(1 - 2\rho)$$
$$- \rho \log \rho - (1-\rho)\log(1-\rho)\}. \quad (10)$$

The equation determining the maximum in Equation 10 is

$$\nu(1-2\rho)+B=\frac{1}{2}\log\left(\frac{1-\rho}{\rho}\right). \tag{11}$$

Substituting $1-2\rho=\eta$ and using the fact that

$$\tanh^{-1}x=\frac{1}{2}\log\left(\frac{1+x}{1-x}\right), \tag{12}$$

Equation 11 becomes

$$\eta=\tanh(\nu\eta+B), \tag{13}$$

which is precisely the mean field equation (Equation 5.21 of Chapter 4). In terms of η satisfying Equation 13, the right-hand side of Equation 10 can be written as

$$\begin{aligned}\nu\left(\frac{1+\eta^2}{2}\right)+\eta B-\left(\frac{1-\eta}{2}\right)\log\left(\frac{1-\eta}{2}\right)-\left(\frac{1+\eta}{2}\right)\log\left(\frac{1+\eta}{2}\right)\\ =\nu\left(\frac{1-\eta^2}{2}\right)-\frac{1}{2}\log\left(\frac{1-\eta^2}{4}\right)\\ =\nu\left(\frac{1-\eta^2}{2}\right)+\log[2\cosh(\nu\eta+B)]. \qquad (14)\end{aligned}$$

We have therefore proved the following

COROLLARY. *If*

$$Z(N)=\sum_{\{\mu\}}\exp\left[\beta\gamma\sum_{i\neq j}v(\gamma|i-j|)\left(\frac{\mu_i+\mu_j}{2}\right)^2+B\sum_{i=1}^{N}\mu_i\right] \tag{15}$$

and $v(x)\geq 0$, *then*

$$\lim_{\gamma\to 0}\lim_{N\to\infty}N^{-1}\log Z(N)=\max_{-\infty<\eta<\infty}\left\{\nu\left(\frac{1-\eta^2}{2}\right)+\log[2\cosh(\nu\eta+B)]\right\}. \tag{16}$$

With the original interaction Equation 1, $\nu/2$ is subtracted from the right-hand side of Equation 16 (see Equations 2 and 5.30 of Chapter 4) and we are left with the Curie–Weiss result (Equation 5.19 of Chapter 4). We now prove the theorem.

PROOF OF THEOREM. Our strategy will be to obtain upper and lower bounds for $Z(N, N_-)$ and show that in the limit $\gamma\to 0$ after the thermodynamic limit $N, N_-\to\infty$ with fixed (spin) density $\rho=N_-/N$, the two bounds

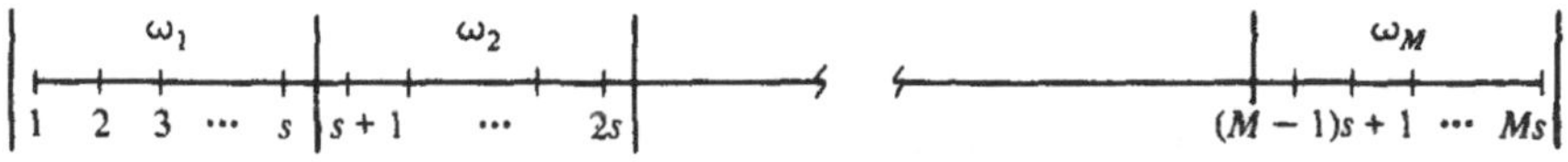

FIGURE C.1. Partitioning of a chain of $N = Ms$ sites into M strips $\omega_1, \ldots, \omega_M$, each containing s sites.

become equal. Let us now obtain an upper bound for $Z(N, N_-)$. The first step is to partition the chain of N sites into M strips $\omega_1, \omega_2, \ldots, \omega_M$ each containing s sites (so that $N = Ms$), as shown in Figure C.1.

We now define

$$Z(N, N_-; N_1, N_2, \ldots, N_M) = \sum_{\{\mu\}}'' \exp\left\{\beta\gamma \sum_{i \neq j} v(\gamma|i - j|)\left(\frac{\mu_i + \mu_j}{2}\right)^2\right\}, \quad (17)$$

where the restricted sum is over configurations $\{\mu\}$ with fixed number $N_i (0 \leq N_i \leq s)$ of down spins in strips ω_i, $i = 1, 2, \ldots, M$, and total number of down spins equal to N_-, i.e.,

$$N_- = \sum_{i=1}^{M} N_i. \quad (18)$$

Obviously the number of up spins ($\mu_i = +1$) in strip ω_i is $s - N_i$ and, from Equations 3 and 17,

$$Z(N, N_-) = \sum_{N_1, \ldots, N_M}' Z(N, N_-; N_1, \ldots, N_M) \quad (19)$$

with the restrictions noted above on the sum in Equation 19. The total number of terms in the sum (Equation 19) does not exceed

$$\frac{(N_- + M - 1)!}{N_-!(M - 1)!} \quad (20)$$

(the number of terms is, in fact, Equation 20 if the restriction $N_i \leq s$ is ignored), so

$$Z(N, N_-) \leq \frac{(N_- + M - 1)!}{N_-!(M - 1)!} \max_{N_1, \ldots, N_M}' Z(N, N_-; N_1, \ldots, N_M). \quad (21)$$

Now when there are N_k down spins (and $s - N_k$ up spins) in strip ω_k and N_l down spins (and $s - N_l$ up spins) in strip ω_l,

$$\sum_{\substack{i \in \omega_k \\ j \in \omega_l}} \left(\frac{\mu_i + \mu_j}{2}\right)^2 = N_k N_l + (s - N_k)(s - N_l) \quad (22)$$

since

$$\left(\frac{\mu_i + \mu_j}{2}\right)^2 = \begin{cases} 1 & \text{if either } \mu_i = \mu_j = -1 \text{ or } \mu_i = \mu_j = +1 \\ 0 & \text{otherwise.} \end{cases} \tag{23}$$

It follows that the summand in Equation 17 is bounded above by

$$\exp\left[\beta\gamma \sum_{i \neq j = 1}^{N} v(\gamma\,|i-j|)\left(\frac{\mu_i + \mu_j}{2}\right)^2\right]$$
$$\leq \exp\left[\beta\gamma \sum_{k,l=1}^{M} v_{\max}(|k-l|)(N_k N_l + (s - N_k)(s - N_l))\right], \tag{24}$$

where

$$v_{\max}(|k-l|) = \max_{\substack{i \in \omega_k \\ j \in \omega_l}} v(\gamma\,|i-j|), \tag{25}$$

i.e., the maximum interaction potential between strips ω_k and ω_l. Since there are

$$A(N_k) = \frac{s!}{N_k!\,(s - N_k)!} \tag{26}$$

ways of arranging N_k down spins in strip ω_k, Equations 17, 21, and 24 give

$$Z(N, N_-) \leq \frac{(N_- + M - 1)!}{N_-!\,(M-1)!} \max_{N_1, \ldots, N_M}{}' \left\{\prod_{k=1}^{M} A(N_k)\right\}$$
$$\times \exp\left\{\beta\gamma \sum_{k,l=1}^{M} v_{\max}(|k-l|)(N_k N_l + (s - N_k)(s - N_l))\right\}. \tag{27}$$

Now

$$N_k N_l \leq \tfrac{1}{2}(N_k^2 + N_l^2)$$

and (28)

$$(s - N_k)(s - N_l) \leq \tfrac{1}{2}[(s - N_k)^2 + (s - N_l)^2],$$

so by symmetry,

$$\sum_{k,l=1}^{M} v_{\max}(|k-l|)(N_k N_l + (s - N_k)(s - N_l)$$
$$\leq \sum_{k=1}^{M} [N_k^2 + (s - N_k)^2] \sum_{l=k}^{M} v_{\max}(|k-l|) \tag{29}$$
$$\leq \sum_{k=1}^{M} [N_k^2 + (s - N_k)^2] \sum_{l=0}^{\infty} v_{\max}(l).$$

From Equations 27 and 29 we then obtain

$$\log Z(N, N_-) \leq \log[(N_- + M - 1)!/N_-!(M-1)!] + \max'_{N_1,\ldots,N_M} \sum_{k=1}^{M} \{\log A(N_k) + \beta\alpha_{\max}[N_k^2 + (s - N_k)^2]\}, \tag{30}$$

where

$$\alpha_{\max} = \gamma \sum_{k=0}^{\infty} v_{\max}(k). \tag{31}$$

The second term on the right-hand side of Equation 30 can be simplified by noting that a function $F(x)$ is bounded above by its convex envelope CE$\{F(x)\}$ [i.e., the minimal convex function which is not less than $F(x)$], as shown in Figure C.2. The convex envelope is a convex function and for such functions $g(x)$,

$$\sum_{i=1}^{M} p_i g(x_i) \leq g\left(\sum_{i=1}^{M} p_i x_i\right) \qquad \text{for } p_i \geq 0 \text{ and } \sum_{i=1}^{M} p_i = 1. \tag{32}$$

In particular, for the function

$$F(N_k) = \log A(N_k) + \beta\alpha_{\max}[N_k^2 + (s - N_k)^2] \tag{33}$$

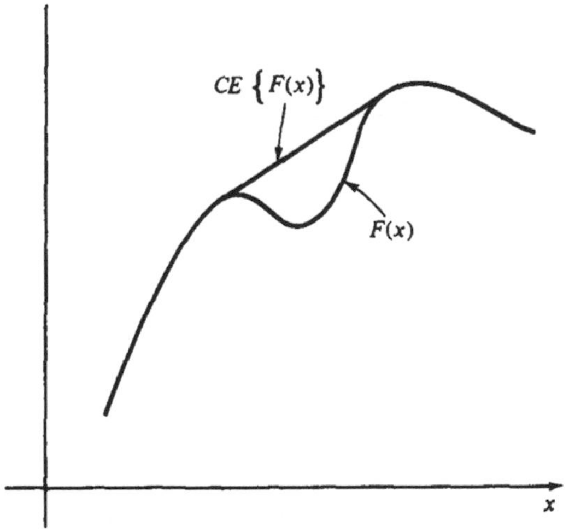

FIGURE C.2. Convex envelope CE[$F(x)$] of a function $F(x)$.

in Equation 30 with $p_i = M^{-1}$ in Equation 32,

$$M^{-1}\sum_{k=1}^{M} F(N_k) \leq M^{-1}\sum_{k=1}^{M} \mathrm{CE}[F(N_k)]$$

$$\leq \mathrm{CE}\left\{F\left(M^{-1}\sum_{k=1}^{M} N_k\right)\right\} \tag{34}$$

$$= \mathrm{CE}\left\{F\left(\frac{N_-}{M}\right)\right\},$$

where in the last step we have used the fact that $\sum_{k=1}^{M} N_k = N_-$ (Equation 18).

We have then, from Equations 30, 33, and 34 that

$$\log Z(N, N_-) \leq \log[(N_- + M - 1)!/N_-!(M-1)!]$$

$$+ M\,\mathrm{CE}\{\log A(\rho s) + \beta s^2 \alpha_{\max}[\rho^2 + (1-\rho)^2]\}, \tag{35}$$

where the "down-spin density" ρ is defined by

$$\rho = \frac{N_-}{N} = \frac{N_-}{Ms}, \qquad \text{i.e., } \rho s = \frac{N_-}{M}. \tag{36}$$

Dividing now by N and taking the limit $N \to \infty$ *with s and ρ fixed* we obtain, using Stirling's formula $\log N! \sim N \log N - N$ as $N \to \infty$ for the first term in Equation 35,

$$f(\rho, s, \gamma) = \lim_{\substack{N, N_- \to \infty \\ \rho, s \text{ fixed}}} N^{-1} \log Z(N, N_-)$$

$$\leq \rho \log\left(1 + \frac{1}{\rho s}\right) + \frac{1}{s}\log(1 + \rho s) \tag{37}$$

$$+ \mathrm{CE}\{s^{-1}\log A(\rho s) + \beta s \alpha_{\max}[\rho^2 + (1-\rho)^2]\}.$$

We now take the limit $\gamma \to 0$. To do this we require that the limit of a convex envelope is the convex envelope of the limit. The proof is trivial. We then only have to evaluate

$$\lim_{\gamma \to 0} s\alpha_{\max} = \lim_{\gamma \to 0} s\gamma \sum_{k=0}^{\infty} v_{\max}(k), \tag{38}$$

where $v_{\max}(k)$, Equation 25, is the maximum of $v(\gamma|i-j|)$ for $i \in \omega_l$ and $j \in \omega_{l+k}$. A typical situation [for monotonic $v(x)$] is shown in Figure C.3.

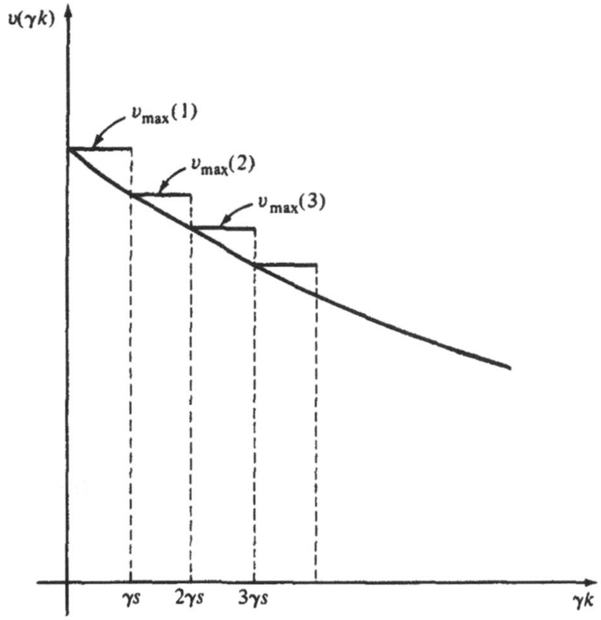

FIGURE C.3. Approximation of an integral by a sum (Equation 39).

and it is obvious that [in general for allowable $v(x)$'s] the sum (Equation 38) approaches the Riemann integral of $v(x)$, i.e.,

$$\begin{aligned}\alpha &= \lim_{\gamma \to 0} s\gamma \sum_{k=0}^{\infty} v_{max}(k) \\ &= \int_0^{\infty} v(x)\, dx < \infty\end{aligned} \tag{39}$$

for arbitrary s. Similarly, if $v_{min}(k) = \min_{\substack{i \in \omega_l \\ j \in \omega_{l+k}}} v(\gamma|i-j|)$, the same reasoning implies that

$$\alpha = \lim_{\gamma \to 0} s\gamma \sum_{k=0}^{\infty} v_{min}(k). \tag{40}$$

We shall need this result for the lower bound.

Finally, we let the (arbitrary) parameter s approach infinity. The first term in Equation 37 vanishes in the limit $s \to \infty$, and recalling the definition equation (26) of $A(\rho s)$, i.e.,

$$A(\rho s) = \frac{s!}{(\rho s)!\,[s(1-\rho)]!},$$

we obtain using Stirling's formula the final result for the upper bound,

$$f(\rho, T) = \lim_{s\to\infty} \lim_{\gamma\to 0} \lim_{\substack{N, N_- \to\infty \\ \rho, s \text{ fixed}}} N^{-1} \log Z(N, N)_- \\ \leq \mathrm{CE}\{\nu[\rho^2 + (1-\rho)^2] - \rho \log \rho - (1-\rho)\log(1-\rho)\}, \tag{41}$$

where $\nu = \beta\alpha$.

To obtain a lower bound we start from Equation 19 and replace the sum on the right-hand side by the particular term with all the N_k's equal; i.e., since $\sum_{k=1}^{M} N_k = N_-$,

$$N_k = \frac{N_-}{M} = \rho s, \qquad k = 1, 2, \ldots, M. \tag{42}$$

With ρs down spins in each strip we have in place of Equation 24,

$$\exp\left[\beta\gamma \sum_{i\neq j=1}^{N} v(\gamma|i-j|)\left(\frac{\mu_i + \mu_j}{2}\right)^2\right] \\ \geq \exp\left\{\beta\gamma s^2[\rho^2 + (1-\rho)^2] \sum_{k,l=1}^{M} v_{\min}(|k-l|)\right\}, \tag{43}$$

where

$$v_{\min}(|k-l|) = \min_{\substack{i \in \omega_k \\ j \in \omega_l}} v(\gamma|i-j|), \tag{44}$$

i.e., the minimum interaction potential between spins in strips ω_k and ω_l. The number of configurations with ρs down spins in each strip is

$$[A(\rho s)]^M = \left[\frac{s!}{(\rho s)!\,[s(1-\rho)]!}\right]^M, \tag{45}$$

so in place of Equation 30 we have the lower bound

$$\log Z(N, N_-) \geq M\left\{\log A(\rho s) + [\rho^2 + (1-\rho)^2]\beta\gamma s^2 M^{-1} \sum_{k,l=1}^{M} v_{\min}(|k-l|)\right\}. \tag{46}$$

Using the obvious result

$$\lim_{M\to\infty} M^{-1} \sum_{k,l=1}^{M} v_{\min}(|k-l|) = \sum_{k=0}^{\infty} v_{\min}(k) = \alpha_{\min}, \tag{47}$$

we obtain after dividing by $N = Ms$ and taking the limit $N \to \infty$ with fixed ρ and s,

$$\lim_{\substack{N, N_- \to \infty \\ \rho, s \text{ fixed}}} N^{-1} \log Z(N, N_-) \geq s^{-1} \log A(\rho s) + [\rho^2 + (1-\rho)^2]\beta s\gamma \sum_{k=0}^{\infty} v_{\min}(k). \tag{48}$$

Now on quite general grounds (see Section 3-7) we know that the limiting function on the left-hand side of Equation 48 is a convex function of ρ. It cannot, therefore, be less than the minimal convex function, which is not less than the right-hand side of Equation 48. In other words,

$$\lim_{\substack{N, N_- \to \infty \\ \rho, s \text{ fixed}}} N^{-1} \log Z(N, N_-) \geq \mathrm{CE}\left\{s^{-1} \log A(\rho s) + [\rho^2 + (1-\rho)^2]\beta s\gamma \sum_{k=0}^{\infty} v_{\min}(k)\right\}. \tag{49}$$

We now take the limit $\gamma \to 0$ followed by $s \to \infty$ and use the results 40 and 41 to obtain

$$\lim_{s \to \infty} \lim_{\gamma \to 0} \lim_{\substack{N, N_- \to \infty \\ \rho, s \text{ fixed}}} N^{-1} \log Z(N, N_-) \geq \mathrm{CE}\{\nu[\rho^2 + (1-\rho)^2] - \rho \log \rho - (1-\rho)\log(1-\rho)\}. \tag{50}$$

This result and Equation 41 implies that

$$f(\rho, T) = \mathrm{CE}\{\nu[\rho^2 + (1-\rho)^2] - \rho \log \rho - (1-\rho) \log(1-\rho)\}, \tag{51}$$

which was to be proved.

Finally a few remarks.

1. We have stated and proved the theorem in one dimension only for reasons of simplicity. It is obvious that the theorem and proof can be straightforwardly generalized to d dimensions with potential $\phi(r) = \gamma^d v(\gamma r)$. The final result (5) is unchanged for $v(r) \geq 0$.

2. The condition $v(r) \geq 0$ is unnecessarily restrictive. Theorems with this condition relaxed have been stated and proved by Lebowitz and Penrose (1966) and more recently by Gates and Penrose (1969).

3. The Kac-type potential $\gamma^d v(\gamma r)$ is also much too restrictive. An interesting case, where the final result is the same, is for potentials of the type

$$\phi(r) = \frac{\varepsilon\alpha}{r^{d+\varepsilon}} \tag{52}$$

with $\varepsilon \to 0$ after the thermodynamic limit. The only difference in the proof occurs in Equations 39 and 40, which are still valid for 52. The reader may amuse himself by trying to find other potentials which give the result 5.

4. It is obviously much easier to prove the corollary directly without first proving the theorem. This is left as an exercise for the reader.

5. Finally, we might mention Lieb's (1966) extension of the Lebowitz–Penrose theorem to quantum-mechanical systems.

APPENDIX D

Algebraic Derivation of the Partition Function for a Two-dimensional Ising Model

We have reduced the problem of evaluating the partition function in Section 5-4 to the problem of finding the largest eigenvalue of the 2^m by 2^m transfer matrix $\mathbf{L}$ with components

$$\mathbf{L}(\sigma, \sigma') = \exp\left(\nu \sum_{k=1}^{m-1} \mu_k \mu_{k+1} + B \sum_{k=1}^{m} \mu_k\right) \exp\left(\nu \sum_{k=1}^{m} \mu_k \mu_k'\right), \tag{1}$$

where $\sigma = (\mu_1, \ldots, \mu_m)$ and $\sigma' = (\mu_1', \ldots, \mu_m')$. The partition function for an m by n lattice with periodicity in rows was shown to be

$$\begin{aligned} Z_{n,m} &= \mathrm{Tr}(\mathbf{L}^n) \\ &= \sum_{j=1}^{2^m} \lambda_j^n, \end{aligned} \tag{2}$$

where the λ_j are the eigenvalues of the matrix $\mathbf{L}$. If, in addition, we impose periodic boundary conditions on columns (so that the lattice is wrapped on a torus), the $m - 1$ in Equation 1 is replaced by m. We will henceforth consider this problem.

In the thermodynamic limit $n \to \infty$ followed by $m \to \infty$ the free energy per spin ψ from Equation 2 is given by

$$\begin{aligned} -\frac{\psi}{kT} &= \lim_{m\to\infty} \lim_{n\to\infty} (mn)^{-1} \log Z_{n,m} \\ &= \lim_{m\to\infty} m^{-1} \log \lambda_1, \end{aligned} \tag{3}$$

where λ_1 is the maximum eigenvalue of $\mathbf{L}$.

To set up an operator algebra for the two-dimensional problem let us return for a moment to the one-dimensional problem in zero field, with transfer matrix (Equation 3.10 of Chapter 5)

$$\mathbf{L} = \begin{pmatrix} e^{\nu} & e^{-\nu} \\ e^{-\nu} & e^{\nu} \end{pmatrix} = \mathbf{I}^2 e^{\nu} + \boldsymbol{\tau}^1 e^{-\nu}, \tag{4}$$

where $\mathbf{I}^2$ is the 2 by 2 unit matrix

$$\mathbf{I}^2 = \begin{pmatrix} 1 & 0 \\ 0 & 1 \end{pmatrix} \tag{5}$$

and $\boldsymbol{\tau}^1$ is defined by

$$\boldsymbol{\tau}^1 = \begin{pmatrix} 0 & 1 \\ 1 & 0 \end{pmatrix}. \tag{6}$$

We now write Equation 4 in the form

$$\mathbf{L} = (2 \sinh 2\nu)^{1/2} \exp(\nu^* \boldsymbol{\tau}^1), \tag{7}$$

where ν^* is defined by

$$\tanh \nu^* = e^{-2\nu}$$

or, equivalently,

$$\sinh 2\nu^* \sinh 2\nu = 1. \tag{8}$$

Equation 7 is easily verified by expanding the exponential and using the fact that $(\boldsymbol{\tau}^1)^2 = \mathbf{I}^2$, i.e.,

$$\begin{aligned} \exp(\nu^* \boldsymbol{\tau}^1) &= \mathbf{I}^2 + (\nu^* \boldsymbol{\tau}^1) + \frac{(\nu^* \boldsymbol{\tau}^1)^2}{2!} + \cdots \\ &= \mathbf{I}^2 \left(1 + \frac{\nu^{*2}}{2!} + \frac{\nu^{*4}}{4!} + \cdots \right) + \boldsymbol{\tau}^1 \left(\nu^* + \frac{\nu^{*3}}{3!} + \frac{\nu^{*5}}{5!} + \cdots \right) \\ &= \mathbf{I}^2 \cosh \nu^* + \boldsymbol{\tau}^1 \sinh \nu^*. \end{aligned} \tag{9}$$

Equations 4 and 7 are then identical in view of Equation 8.

To simplify the two-dimensional transfer matrix, Equation 1, consider first the matrix $\mathbf{V}_1'$ defined by

$$\begin{aligned} V_1'(\sigma, \sigma') &= \exp\left(\nu \sum_{k=1}^{m} \mu_k \mu_k'\right) \\ &= \prod_{k=1}^{m} \exp(\nu \mu_k \mu_k'). \end{aligned} \tag{10}$$

This matrix is simply a direct product of the matrices Equation 7, i.e.,

$$\mathbf{V}_1' = (2 \sinh 2\nu)^{m/2} \exp(\nu^* \tau^1) \otimes \exp(\nu^* \tau^1) \otimes \cdots \otimes \exp(\nu^* \tau^1). \tag{11}$$

At this stage a few words about direct products are in order.

Consider two n by n matrices A and B with matrix elements $A_{i,j}$ and $B_{i',j'}$. The direct product matrix $A \otimes B$ is an n^2 by n^2 matrix with elements

$$(A \otimes B)_{ii',jj'} = A_{i,j} B_{i',j'}. \tag{12}$$

For example, if

$$A = \begin{pmatrix} A_{11} & A_{12} \\ A_{21} & A_{22} \end{pmatrix} \quad \text{and} \quad B = \begin{pmatrix} B_{11} & B_{12} \\ B_{21} & B_{22} \end{pmatrix},$$

then

$$A \otimes B = \left(\begin{array}{c|c} A_{11}B & A_{12}B \\ \hline A_{21}B & A_{22}B \end{array}\right)$$

$$= \begin{pmatrix} A_{11}B_{11} & A_{11}B_{12} & A_{12}B_{11} & A_{12}B_{12} \\ A_{11}B_{21} & A_{11}B_{22} & A_{12}B_{21} & A_{12}B_{22} \\ A_{21}B_{11} & A_{21}B_{12} & A_{22}B_{11} & A_{22}B_{12} \\ A_{21}B_{21} & A_{21}B_{22} & A_{22}B_{21} & A_{22}B_{22} \end{pmatrix}.$$

Equation 11 follows straightforwardly from the definition Equation 12.

We record the following properties of direct products for later reference.

1. Under ordinary matrix multiplication

$$(A \otimes B)(C \otimes D) = AC \otimes BD. \tag{13}$$

By definition, the (ii', jj') element of the left-hand side of Equation 13 is

$$\sum_{k,k'} (A \otimes B)_{ii',kk'} (C \otimes D)_{kk',jj'}$$

$$= \sum_k A_{ik} C_{kj} \sum_{k'} B_{i'k'} D_{k'j'}$$

$$= (AC)_{ij} (BD)_{i'j'},$$

which proves Equation 13.

2. If the matrix A is diagonalized by S and B by T, then $A \otimes B$ is diagonalized by $S \otimes T$. This follows from Equation 13 since

$$(S \otimes T)^{-1} = S^{-1} \otimes T^{-1} \tag{14}$$

and

$$(S^{-1} \otimes T^{-1})(A \otimes B)(S \otimes T) = S^{-1}AS \otimes T^{-1}BT. \tag{15}$$

3. As a consequence of Property 2, the eigenvalues of $A \otimes B$ are $a_i b_j$, $i, j = 1, 2, \ldots, n$, where a_i are the eigenvalues of A and b_j the eigenvalues of B.

4. It follows from property 3 that

$$\mathrm{Tr}(A \otimes B) = (\mathrm{Tr}\, A)(\mathrm{Tr}\, B). \tag{16}$$

From property 1 (Equation 13) it follows that $\mathbf{V}_1'$ (Equation 11) can be written in the form

$$\mathbf{V}_1' = (2 \sinh 2v)^{m/2} \exp\left(v^* \sum_{k=1}^{m} \boldsymbol{\tau}_k^1\right), \tag{17}$$

where

$$\boldsymbol{\tau}_k^1 = \mathbf{I}^2 \otimes \mathbf{I}^2 \otimes \cdots \otimes \mathbf{I}^2 \otimes \boldsymbol{\tau}^1 \otimes \mathbf{I}^2 \otimes \cdots \otimes \mathbf{I}^2, \qquad k = 1, 2, \ldots, m, \tag{18}$$

with m terms in the direct product and $\boldsymbol{\tau}^1$ (Equation 6) in the kth place.

Returning now to Equation 1 we write the matrix $\mathbf{L}$ in the form

$$\mathbf{L} = \mathbf{V}_1'\mathbf{V}_2\mathbf{V}_3, \tag{19}$$

where $\mathbf{V}_1'$ is given by Equations 10 and 17, and, from Equation 1,

$$V_2(\sigma, \sigma') = \delta_{\mu_1, \mu_1'}, \delta_{\mu_2, \mu_2'} \cdots \delta_{\mu_m, \mu_m'} \prod_{k=1}^{m} \exp(v\mu_k \mu_{k+1}) \tag{20}$$

and

$$V_3(\sigma, \sigma') = \delta_{\mu_1, \mu_1'} \delta_{\mu_2, \mu_2'} \cdots \delta_{\mu_m, \mu_m'} \prod_{k=1}^{m} \exp(B\mu_k). \tag{21}$$

The matrices $\mathbf{V}_2$ and $\mathbf{V}_3$ are therefore diagonal in the representation Equation 17 for $\mathbf{V}_1'$, and it is obvious from the above that we can write

$$\mathbf{V}_2 = \exp\left(v \sum_{k=1}^{m} \boldsymbol{\tau}_k^3 \boldsymbol{\tau}_{k+1}^3\right) \tag{22}$$

and

$$\mathbf{V}_3 = \exp\left(B \sum_{k=1}^{m} \boldsymbol{\tau}_k^3\right), \tag{23}$$

where

$$\boldsymbol{\tau}_k^3 = \mathbf{I}^2 \otimes \mathbf{I}^2 \otimes \cdots \otimes \mathbf{I}^2 \otimes \boldsymbol{\tau}^3 \otimes \mathbf{I}^2 \otimes \cdots \otimes \mathbf{I}^2 \tag{24}$$

with the matrix

$$\boldsymbol{\tau}^3 = \begin{pmatrix} 1 & 0 \\ 0 & -1 \end{pmatrix} \tag{25}$$

in the kth place (the $+1$ corresponding to $\mu = +1$ and the -1 to $\mu = -1$).

We then write the transfer matrix $\mathbf{L}$ as

$$\mathbf{L} = (2 \sinh 2\nu)^{m/2} \mathbf{V}_1 \mathbf{V}_2 \mathbf{V}_3, \tag{26}$$

where

$$\mathbf{V}_1 = \exp\left(\nu^* \sum_{k=1}^{m} \boldsymbol{\tau}_k^1\right), \tag{27}$$

and $\mathbf{V}_2$ and $\mathbf{V}_3$ are defined, respectively, by Equations 22 and 23. From this point on we shall consider the zero-field case only; i.e., from Equation 23 $\mathbf{V}_3 = \mathbf{I}^{2^m}$, the 2^m-dimensional identity matrix.

From Equations 2 and 26 (with $B = 0$) the partition function is given by

$$\begin{aligned} Z_{n,m} &= (2 \sinh 2\nu)^{mn/2} \operatorname{Tr}([\mathbf{V}_1 \mathbf{V}_2]^n) \\ &= (2 \sinh 2\nu)^{mn/2} \sum_{j=1}^{2^m} \Lambda_j^n, \end{aligned} \tag{28}$$

where Λ_j are the eigenvalues of the matrix $\mathbf{V}_1 \mathbf{V}_2$.

The matrices $\boldsymbol{\tau}_k^1$ and $\boldsymbol{\tau}_k^3$ defined above are familiar in quantum mechanics: They are two of the Pauli spin matrices, the third $\boldsymbol{\tau}_k^2$ being

$$\boldsymbol{\tau}_k^2 = -i\boldsymbol{\tau}_k^3 \boldsymbol{\tau}_k^1 = \mathbf{I}^2 \otimes \cdots \otimes \mathbf{I}^2 \otimes \boldsymbol{\tau}^2 \otimes \mathbf{I}^2 \otimes \cdots \otimes \mathbf{I}^2, \tag{29}$$

where $i^2 = -1$ and

$$\boldsymbol{\tau}^2 = -i\boldsymbol{\tau}^3 \boldsymbol{\tau}^1 = \begin{pmatrix} 0 & -i \\ i & 0 \end{pmatrix}. \tag{30}$$

The following properties of the Pauli matrices, which we shall need in a moment, are easily verified:

$$\boldsymbol{\tau}_k^\alpha \boldsymbol{\tau}_k^\beta = i\boldsymbol{\tau}_k^\gamma \qquad \text{for } (\alpha\beta\gamma) \text{, a cyclic permutation of } (123) \tag{31}$$

$$(\boldsymbol{\tau}_k^\alpha)^2 = \mathbf{I}^{2^m}, \qquad \text{the } 2^m \text{ by } 2^m \text{ unit matrix, for } \alpha = 1, 2, 3 \text{ and } k = 1, 2, \cdots, m. \tag{32}$$

$$\boldsymbol{\tau}_k^\alpha \boldsymbol{\tau}_k^\beta + \boldsymbol{\tau}_k^\beta \boldsymbol{\tau}_k^\alpha = \mathbf{O}^{2^m}, \qquad \text{the } 2^m \text{ by } 2^m \text{ null matrix, for } \alpha = 1, 2, 3, \ \beta = 1, 2, 3, \text{ and } k = 1, 2, \ldots, m, \ (\alpha \neq \beta), \tag{33}$$

and

$$\boldsymbol{\tau}_k^\alpha \boldsymbol{\tau}_l^\beta = \boldsymbol{\tau}_l^\beta \boldsymbol{\tau}_k^\alpha \qquad \text{for } k \neq l = 1, 2, \ldots, m \text{ and } \alpha, \beta = 1, 2, 3. \tag{34}$$

The next step in the evaluation of $Z_{n,m}$ is to define matrices $\mathbf{P}_k$ and $\mathbf{Q}_k$, $k = 1, 2, \ldots, m$ by

$$\mathbf{P}_k = \boldsymbol{\tau}_1^1 \boldsymbol{\tau}_2^1 \cdots \boldsymbol{\tau}_{k-1}^1 \boldsymbol{\tau}_k^3 \tag{35}$$

and

$$\mathbf{Q}_k = \boldsymbol{\tau}_1^1 \boldsymbol{\tau}_2^1 \cdots \boldsymbol{\tau}_{k-1}^1 \boldsymbol{\tau}_k^2. \tag{36}$$

The reasons for considering these matrices will become clear in a moment.

In view of the relations 31–34 for the spin matrices $\boldsymbol{\tau}_k^\alpha$, we have that

$$\mathbf{P}_k \mathbf{Q}_k = \boldsymbol{\tau}_k^3 \boldsymbol{\tau}_k^2 = -i\boldsymbol{\tau}_k^1, \tag{37}$$

so the matrix $\mathbf{V}_1$, Equation 27, can be written as

$$\mathbf{V}_1 = \exp\left(iv^* \sum_{k=1}^{m} \mathbf{P}_k \mathbf{Q}_k\right). \tag{38}$$

Similarly,

$$\begin{aligned} \mathbf{P}_{k+1}\mathbf{Q}_k &= \boldsymbol{\tau}_{k+1}^3 \boldsymbol{\tau}_k^1 \boldsymbol{\tau}_k^2 \\ &= i\boldsymbol{\tau}_{k+1}^3 \boldsymbol{\tau}_k^3, \qquad \text{for } k = 1, 2, \ldots, m-1. \end{aligned} \tag{39}$$

The boundary term $\boldsymbol{\tau}_1^3 \boldsymbol{\tau}_m^3$ in the matrix $\mathbf{V}_2$ is given by

$$\boldsymbol{\tau}_1^3 \boldsymbol{\tau}_m^3 = i\mathbf{P}_1 \mathbf{Q}_m \mathbf{U}, \tag{40}$$

where

$$\mathbf{U} = \boldsymbol{\tau}_1^1 \boldsymbol{\tau}_2^1 \cdots \boldsymbol{\tau}_m^1. \tag{41}$$

The matrix $\mathbf{V}_2$ can then be written in the form

$$\mathbf{V}_2 = \exp(iv\mathbf{P}_1\mathbf{Q}_m\mathbf{U})\exp\left(-iv\sum_{k=1}^{m-1}\mathbf{P}_{k+1}\mathbf{Q}_k\right). \tag{42}$$

The boundary term in Equation 42 is obviously a nuisance, but, as we will see in a moment, it simplifies the evaluation of $Z_{n,m}$. To write $\mathbf{V}_2$ in a more convenient form we use the results

$$\mathbf{U}^2 = \mathbf{I}^{2^m} \qquad \text{and} \qquad (i\mathbf{P}_1\mathbf{Q}_m\mathbf{U})^2 = \mathbf{I}^{2^m}, \tag{43}$$

which follow directly from Equations 40 and 41, to write (see Equation 9)

$$\begin{aligned}\exp(iv\mathbf{P}_1\mathbf{Q}_m\mathbf{U}) &= \mathbf{I}^{2^m}\cosh v + i\mathbf{P}_1\mathbf{Q}_m\mathbf{U}\sinh v\\ &= [\tfrac{1}{2}(\mathbf{I}^{2^m}+\mathbf{U}) + \tfrac{1}{2}(\mathbf{I}^{2^m}-\mathbf{U})](\cosh v + i\mathbf{P}_1\mathbf{Q}_m\mathbf{U}\sinh v)\\ &= \tfrac{1}{2}(\mathbf{I}^{2^m}+\mathbf{U})(\cosh v + i\mathbf{P}_1\mathbf{Q}_m\sinh v)\\ &\quad + \tfrac{1}{2}(\mathbf{I}^{2^m}-\mathbf{U})(\cosh v - i\mathbf{P}_1\mathbf{Q}_m\sinh v)\\ &= \tfrac{1}{2}(\mathbf{I}^{2^m}+\mathbf{U})\exp(iv\mathbf{P}_1\mathbf{Q}_m) + \tfrac{1}{2}(\mathbf{I}^{2^m}-\mathbf{U})\exp(-iv\mathbf{P}_1\mathbf{Q}_m).\end{aligned} \tag{44}$$

The matrix $\mathbf{V} = \mathbf{V}_2\mathbf{V}_1$ can therefore be written in the form

$$\mathbf{V} = \tfrac{1}{2}(\mathbf{I}^{2^m}+\mathbf{U})\mathbf{V}_+ + \tfrac{1}{2}(\mathbf{I}^{2^m}-\mathbf{U})\mathbf{V}_-, \tag{45}$$

where

$$\mathbf{V}_\pm = \exp\left(-iv\sum_{k=1}^{m}\mathbf{P}_{k+1}\mathbf{Q}_k\right)\exp\left(iv^*\sum_{k=1}^{m}\mathbf{P}_k\mathbf{Q}_k\right), \tag{46}$$

with the convention that

$$\mathbf{P}_{m+1} = \mp\mathbf{P}_1 \qquad \text{for } \mathbf{V}_\pm\,. \tag{47}$$

The reader may wonder why we transformed $\mathbf{V}$ from one seemingly complicated form, Equation 26, to another, Equation 45, so perhaps at this point a little history and motivation is not out of place.

The transformation to the **P** and **Q** operators was first made by Kaufmann (1949a) in her simplification of Onsager's original derivation. The form 45 accomplishes two things. First, if one defines operators $\mathbf{\Gamma}_k$ by

$$\left.\begin{aligned}\mathbf{\Gamma}_{2r-1} &= \mathbf{P}_r\\ \mathbf{\Gamma}_{2r} &= \mathbf{Q}_r\end{aligned}\right\}\text{for } r = 1, 2, \ldots, m, \tag{48}$$

we see from the definitions, Equations 35 and 36, of $\mathbf{P}_r$ and $\mathbf{Q}_r$, that the $\mathbf{\Gamma}$ operators satisfy the anticommutation relations

$$\mathbf{\Gamma}_k\mathbf{\Gamma}_l + \mathbf{\Gamma}_l\mathbf{\Gamma}_k = 2\delta_{k,l}\mathbf{I}^{2^m}, \qquad k, l = 1, 2, \ldots, 2m. \tag{49}$$

Equations 48, 35, and 36 are particular matrix representations of the operator algebra defined by Equation 49. General properties of the algebra (Equation 49) were studied many years ago by Brauer and Weyl [see Murnaghan (B1938)]. The basic result which was exploited by Kaufman is that certain similarity transformations on the $\mathbf{\Gamma}$ operators form 2^m-dimensional representations of the $2m$-dimensional rotation group. This fact reduces the dimensionality, and hence complexity, of the problem considerably. It turns out, in fact, that $\mathbf{V}_+$ and $\mathbf{V}_-$ (but *not* $\mathbf{V}$) are themselves spin representations of

$2m$-dimensional rotations. The second simplification in Equation 45 is then the splitting of $\mathbf{V}$ into $\mathbf{V}_+$ and $\mathbf{V}_-$ (this was a difficulty with Onsager's original derivation). Note that since $\mathbf{U}^2 = \mathbf{I}^{2^m}$, $\mathbf{P}_\pm = \frac{1}{2}(\mathbf{I}^{2^m} \pm \mathbf{U})$ are projection operators (i.e., $\mathbf{P}_\pm^2 = \mathbf{P}_\pm$ and $\mathbf{P}_\pm$ are self-adjoint), $\mathbf{P}_\pm$ onto the *orthogonal* even and odd subspaces, respectively (since $\mathbf{P}_+\mathbf{P}_- = \mathbf{O}$). We shall not pursue this line of argument any further here. Instead we now present a simpler argument based on the work of Schultz *et al.* (1964) and Thompson (1965).

The trick is to consider operators $\mathbf{a}_k$ and their Hermitian adjoints $\mathbf{a}_k^\dagger$, instead of $\mathbf{P}_k$ and $\mathbf{Q}_k$ (Equations 35 and 36), defined by

$$\begin{aligned} \mathbf{a}_k + \mathbf{a}_k^\dagger &= \boldsymbol{\tau}_1^1\boldsymbol{\tau}_2^1 \cdots \boldsymbol{\tau}_{k-1}^1\boldsymbol{\tau}_k^2 = \mathbf{Q}_k, \\ \mathbf{a}_k - \mathbf{a}_k^\dagger &= i\boldsymbol{\tau}_1^1\boldsymbol{\tau}_2^1 \cdots \boldsymbol{\tau}_{k-1}^1\boldsymbol{\tau}_k^3 = i\mathbf{P}_k. \end{aligned} \tag{50}$$

From the properties 31–34 of the $\boldsymbol{\tau}$ matrices, the $\mathbf{a}$ operators are seen to have the anticommutation relations

$$\begin{aligned} \mathbf{a}_k\mathbf{a}_{k'}^\dagger + \mathbf{a}_{k'}^\dagger\mathbf{a}_k &= \mathbf{I}^{2^m}\,\delta_{k,k'}, \\ \mathbf{a}_k\mathbf{a}_{k'} + \mathbf{a}_{k'}\mathbf{a}_k &= \mathbf{O}^{2^m}, \end{aligned} \tag{51}$$

which are the well-known (in physics) relations for Fermi creation ($\mathbf{a}_k^\dagger$) and annihilation ($\mathbf{a}_k$) operators.

In terms of these operators, $\mathbf{V}_\pm$ (Equation 46) and $\mathbf{U}$ (Equation 41) are given by

$$\mathbf{V}_\pm = \exp\left[\nu \sum_{k=1}^{m} (\mathbf{a}_{k+1}^\dagger - \mathbf{a}_{k+1})(\mathbf{a}_k^\dagger + \mathbf{a}_k)\right] \exp\left[-2\nu^* \sum_{k=1}^{m} (\mathbf{a}_k^\dagger\mathbf{a}_k - \tfrac{1}{2})\right], \tag{52}$$

and, from Equations 37 and 41,

$$\begin{aligned} \mathbf{U} &= \prod_{k=1}^{m} \boldsymbol{\tau}_k^1 \\ &= (-i)^m \prod_{k=1}^{m} \exp\left(\frac{\pi \mathbf{P}_k\mathbf{Q}_k}{2}\right) \\ &= (-i)^m \exp\left[i\pi \sum_{k=1}^{m} (\mathbf{a}_k^\dagger\mathbf{a}_k - \tfrac{1}{2})\right], \end{aligned} \tag{53}$$

where use has been made of the relations 51. For $\mathbf{V}_+$ (Equation 52) we have the anticyclic boundary conditions (see Equation 47)

$$\mathbf{a}_{m+1} = -\mathbf{a}_1 \tag{54}$$

and for $\mathbf{V}_-$ we have the cyclic boundary conditions

$$\mathbf{a}_{m+1} = \mathbf{a}_1. \tag{55}$$

Schultz et al. actually found all the eigenvalues of $\mathbf{V}_+$ and $\mathbf{V}_-$ (and hence $\mathbf{V}$). Only the largest eigenvalue is needed to evaluate the partition function, but the other eigenvalues are important for such things as correlation functions (see Section 5-6). To evaluate the trace Equation 28, however, it turns out that one does not need to know the eigenvalues at all; for a special representation of the operators the trace can be evaluated directly.

The first point to note is that

$$\mathbf{V}^n = \tfrac{1}{2}(\mathbf{I}^{2^m} + \mathbf{U})\mathbf{V}_+^n + \tfrac{1}{2}(\mathbf{I}^{2^m} - \mathbf{U})\mathbf{V}_-^n. \tag{56}$$

This is because $\tfrac{1}{2}(\mathbf{I}^{2^m} \pm \mathbf{U})$ are mutually orthogonal projection operators which commute with $\mathbf{V}_+$ and $\mathbf{V}_-$, respectively. The partition function is given by (Equation 28)

$$Z_{n,m} = (2 \sinh 2v)^{mn/2} \operatorname{Tr}(\mathbf{V}^n), \tag{57}$$

which from Equation 56 is a sum of four terms. Let us consider first $\operatorname{Tr}(\mathbf{V}_+^n)$.

To simplify $\mathbf{V}_+$ we first transform to "running wave operators" $\boldsymbol{\eta}_q$, defined by

$$\mathbf{a}_k = m^{-1/2} e^{i\pi/4} \sum_q e^{iqk} \boldsymbol{\eta}_q, \tag{58}$$

where the anticyclic condition Equation 54 requires that

$$q = \pm(2j - 1)\pi/m, \qquad j = 1, 2, \ldots, m/2, \tag{59}$$

in Equation 58, and we have assumed for convenience that m is even. [The cyclic condition (55) requires that $q = 0, \pi$, or $\pm 2j\pi/m, j = 1, 2, \ldots, m/2 - 1$.] The factor $e^{i\pi/4}$ in Equation 58 is included simply to make the final coefficients real. The beauty of the Fermi operators is now clear when one notes that the transformation (58) preserves the Fermi anticommutation relations, Equation 51 (i.e., the $\boldsymbol{\eta}_q$ operators are also Fermi operators, as can be easily verified from Equations 51 and 58), and also factors $\mathbf{V}_+$. Thus substitution of Equation 58 into Equation 52 gives

$$\mathbf{V}_+ = \prod_{0<q<\pi} \mathbf{V}_q, \tag{60}$$

where in terms of the operators $\mathbf{\Sigma}_q^\alpha$, $\alpha = 1, 2, 3$, defined by

$$\begin{aligned}\mathbf{\Sigma}_q^1 &= \boldsymbol{\eta}_{-q}\boldsymbol{\eta}_q + \boldsymbol{\eta}_q^\dagger\boldsymbol{\eta}_{-q}^\dagger,\\ \mathbf{\Sigma}_q^2 &= i(\boldsymbol{\eta}_{-q}\boldsymbol{\eta}_q - \boldsymbol{\eta}_q^\dagger\boldsymbol{\eta}_{-q}^\dagger),\\ \mathbf{\Sigma}_q^3 &= \boldsymbol{\eta}_q^\dagger\boldsymbol{\eta}_q + \boldsymbol{\eta}_{-q}^\dagger\boldsymbol{\eta}_{-q} - \mathbf{I}^{2m},\end{aligned} \tag{61}$$

$$\mathbf{V}_q = \exp[2\nu(\cos q\mathbf{\Sigma}_q^3 - \sin q\mathbf{\Sigma}_q^1)]\exp(-2\nu^*\mathbf{\Sigma}_q^3), \tag{62}$$

and q in the product Equation 60 is $(2j-1)\pi/m$, $j = 1, 2, \ldots, m/2$.

We shall make use in a moment of the following commutation relations for the $\mathbf{\Sigma}_q^\alpha$ operators:

$$\mathbf{\Sigma}_q^\alpha\mathbf{\Sigma}_{q'}^\beta - \mathbf{\Sigma}_{q'}^\beta\mathbf{\Sigma}_q^\alpha = 2i\,\delta_{q,q'}\mathbf{\Sigma}_q^\gamma \qquad \text{for } (\alpha\beta\gamma) \text{ cyclic } (123) \tag{63}$$

and

$$\mathbf{\Sigma}_q^\alpha\mathbf{\Sigma}_q^\beta + \mathbf{\Sigma}_q^\beta\mathbf{\Sigma}_q^\alpha = \mathbf{O}^{2m} \qquad \text{for } \alpha \neq \beta = 1, 2, 3.$$

These follow directly from Equation 61 and the Fermi anticommutation relations for the $\boldsymbol{\eta}_q$ operators.

We remark in passing that the representation (62) was originally obtained by Onsager, using a completely different method. He did not use the Fermi representation Equation 61 of the $\mathbf{\Sigma}$ operators, which as we will see in a moment, simplifies the remainder of the calculation considerably.

For the purpose of computing the trace of $\mathbf{V}_+^n$, it is convenient to use the "occupation-number representation" (in which $\mathbf{\Sigma}_q^3$ is diagonal) rather than the complete diagonal representation of the $\mathbf{V}_q$'s [the reader is referred to Schultz et al. (1964) for this representation].

Since the eigenvalues of $\boldsymbol{\eta}_q^\dagger\boldsymbol{\eta}_q$ and $\boldsymbol{\eta}_{-q}^\dagger\boldsymbol{\eta}_{-q}$ are 0 and 1 with equal degeneracy (this follows from the anticommutation relations), it follows from Equation 61 that the eigenvalues of $\mathbf{\Sigma}_q^3$ are 1, -1, 0, and 0 with equal degeneracy. A representation of the $\mathbf{\Sigma}_q$ operators in which $\mathbf{\Sigma}_q^3$ is diagonal, is therefore, from Equation 63,

$$\mathbf{\Sigma}_q^\alpha = \mathbf{I}^4 \otimes \mathbf{I}^4 \otimes \cdots \otimes \mathbf{I}^4 \otimes \boldsymbol{\sigma}^\alpha\mathbf{r}^+ \otimes \mathbf{I}^4 \otimes \cdots \otimes \mathbf{I}^4, \tag{64}$$

where there are $m/2$ terms in the direct product with $\boldsymbol{\sigma}^\alpha\mathbf{r}^+$ in the jth place $[q = (2j-1)\pi/m]$, $\mathbf{I}^4$ is the 4 by 4 unit matrix, and $\boldsymbol{\sigma}^\alpha$ and $\mathbf{r}^+$ are defined, respectively, by

$$\boldsymbol{\sigma}^\alpha = \begin{pmatrix}\boldsymbol{\tau}^\alpha & \mathbf{0}^2\\ \mathbf{0}^2 & \boldsymbol{\tau}^\alpha\end{pmatrix}, \qquad \mathbf{r}^+ = \begin{pmatrix}\mathbf{I}^2 & \mathbf{0}^2\\ \mathbf{0}^2 & \mathbf{0}^2\end{pmatrix}, \tag{65}$$

where $\mathbf{I}^2$ and $\mathbf{0}^2$ are the 2 by 2 unit and null matrices, respectively, and the $\boldsymbol{\tau}^\alpha$, $\alpha = 1, 2, 3$, are the Pauli matrices defined by (Equations 6, 25, and 30)

$$\boldsymbol{\tau}^1 = \begin{pmatrix} 0 & 1 \\ 1 & 0 \end{pmatrix}, \quad \boldsymbol{\tau}^2 = \begin{pmatrix} 0 & -i \\ i & 0 \end{pmatrix}, \quad \boldsymbol{\tau}^3 = \begin{pmatrix} 1 & 0 \\ 0 & -1 \end{pmatrix}. \tag{66}$$

Now using the multiplication property (Equation 13) of direct products we can write, using Equations 60, 62, and 64,

$$\mathbf{V}_+^n = \bigotimes_{j=1}^{m/2} \mathbf{V}_{2j-1}^n, \tag{67}$$

where

$$\mathbf{V}_l = \exp\left\{2\nu\left[\cos\left(\frac{l\pi}{m}\right)\boldsymbol{\sigma}_+^3 - \sin\left(\frac{l\pi}{m}\right)\boldsymbol{\sigma}_+^1\right]\right\}\exp(-2\nu^*\boldsymbol{\sigma}_+^3), \tag{68}$$

and $\boldsymbol{\sigma}_+^\alpha$ is defined by

$$\boldsymbol{\sigma}_+^\alpha = \boldsymbol{\sigma}^\alpha \mathbf{r}^+. \tag{69}$$

If we then use the trace property (Equation 16) of direct products, we have from Equation 67 that

$$\mathrm{Tr}(\mathbf{V}_+^n) = \prod_{j=1}^{m/2} \mathrm{Tr}(\mathbf{V}_{2j-1}^n). \tag{70}$$

The problem has now been reduced to the evaluation of traces of 4 by 4 matrices with relatively simple structure. Let us consider $\mathbf{V}_l$. Since the matrices $\boldsymbol{\sigma}_+^\alpha$ satisfy the commutation relations (see Equations 31–34)

$$\boldsymbol{\sigma}_+^\alpha \boldsymbol{\sigma}_+^\beta - \boldsymbol{\sigma}_+^\beta \boldsymbol{\sigma}_+^\alpha = 2i\boldsymbol{\sigma}_+^\gamma \qquad \text{for } (\alpha\beta\gamma) \text{ cyclic } (123), \tag{71}$$

$\mathbf{V}_l$ can be written in the form

$$\mathbf{V}_l = \exp\left(\sum_{\alpha=1}^{3} c_l^\alpha \boldsymbol{\sigma}_+^\alpha\right), \tag{72}$$

where c_l^α are numbers to be determined. Equation 72 is a consequence of the fact that $\exp(A)\exp(B)$ can be written in the form $\exp(C)$, where C is a sum of commutators of A and B (the Baker–Hausdorf expansion). In view of the commutation relations (71) we deduce that $\mathbf{V}_l$ (Equation 68) can be written in the form of Equation 72. If we then use the anticommutation relations

$$\boldsymbol{\sigma}_+^\alpha \boldsymbol{\sigma}_+^\beta + \boldsymbol{\sigma}_+^\beta \boldsymbol{\sigma}_+^\alpha = 2\delta_{\alpha,\beta}\mathbf{r}^+, \qquad \alpha, \beta = 1, 2, 3, \tag{73}$$

which follow from Equations 65, 32, and 33, expansion of the exponential in Equation 72 gives

$$\mathbf{V}_l = \mathbf{r}^- + \mathbf{r}^+ \cosh \gamma_l + \left(\sum_{\alpha=1}^{3} c_l^\alpha \boldsymbol{\sigma}_+^\alpha\right) \frac{\sinh \gamma_l}{\gamma_l}, \tag{74}$$

where

$$\mathbf{r}^- = \mathbf{I}^4 - \mathbf{r}^+ \tag{75}$$

and γ_l is defined by

$$\gamma_l^2 = \sum_{\alpha=1}^{3} (c_l^\alpha)^2. \tag{76}$$

If we also expand the exponentials in Equation 68 we obtain

$$\begin{aligned} \mathbf{V}_l = \mathbf{r}^- &+ \mathbf{r}^+ \left[\cosh 2v \cosh 2v^* - \sinh 2v \sinh 2v^* \cos\left(\frac{l\pi}{m}\right)\right] \\ &- \boldsymbol{\sigma}_+^1 \left[\sinh 2v \cosh 2v^* \sin\left(\frac{l\pi}{m}\right)\right] \\ &+ \boldsymbol{\sigma}_+^2 \left[\sinh 2v^* \sinh 2v \sin\left(\frac{l\pi}{m}\right)\right] \\ &+ \boldsymbol{\sigma}_+^3 \left[\sinh 2v \cosh 2v^* \cos\left(\frac{l\pi}{m}\right) - \sinh 2v^* \cosh 2v\right]. \end{aligned} \tag{77}$$

This expression must be identical with Equation 74, so equating coefficients of $\mathbf{r}^+$ we obtain, in view of the definition (Equation 8) of v^*,

$$\begin{aligned} \cosh \gamma_l &= \cosh 2v \cosh 2v^* - \sinh 2v \sinh 2v^* \cos\left(\frac{l\pi}{m}\right) \\ &= \cosh 2v \cosh 2v^* - \cos\left(\frac{l\pi}{m}\right). \end{aligned} \tag{78}$$

It turns out that this is the only relation needed to evaluate the trace equation (70). The steps needed are as follows. From Equations 72 and 74 we have

$$\begin{aligned} \mathbf{V}_l^n &= \exp\left[\sum_{\alpha=1}^{3} (n c_l^\alpha) \boldsymbol{\sigma}_+^\alpha\right] \\ &= \mathbf{r}^- + \mathbf{r}^+ \cosh(n\gamma_l) + \left(\sum_{\alpha=1}^{3} c_l^\alpha \boldsymbol{\sigma}_+^\alpha\right) \frac{\sinh n\gamma_l}{\gamma_l}, \end{aligned} \tag{79}$$

and recalling Equations 65 and 75 for $\mathbf{r}^+$ and $\mathbf{r}^-$, and noting that $\mathrm{Tr}(\boldsymbol{\sigma}_+^a) = 0$, we have

$$\mathrm{Tr}(\mathbf{V}_l^n) = 2[1 + \cosh(n\gamma_l)] = 4\cosh^2\left(\frac{n\gamma_l}{2}\right). \tag{80}$$

This completes the evaluation of the first term in $\mathrm{Tr}(\mathbf{V}^n)$ (Equation 56); i.e., from Equation 70 and 80,

$$\mathrm{Tr}(\mathbf{V}_+^n) = \prod_{j=1}^{m/2} 4\cosh^2\left(\frac{n\gamma_{2j-1}}{2}\right).. \tag{81}$$

To evaluate $\mathrm{Tr}(\mathbf{U}\mathbf{V}_+^n)$ in Equation 57 we use the representation 64 to express $\mathbf{U}$ (Equation 53) in the form

$$\mathbf{U} = \bigotimes_{j=1}^{m/2} (-e^{i\pi\sigma_+^3}) = \bigotimes_{j=1}^{m/2} (\mathbf{r}^+ - \mathbf{r}^-). \tag{82}$$

From Equation 79 and the properties of direct products it follows that

$$\begin{aligned}\mathrm{Tr}(\mathbf{U}\mathbf{V}_+^n) &= \prod_{j=1}^{m/2} \mathrm{Tr}[(\mathbf{r}^+ - \mathbf{r}^-)\mathbf{V}_{2j-1}^n] \\ &= \prod_{j=1}^{m/2} 2[-1 + \cosh(n\gamma_{2j-1})] \\ &= \prod_{j=1}^{m/2} 4\sinh^2\left(\frac{n\gamma_{2j-1}}{2}\right),\end{aligned} \tag{83}$$

where γ_l is defined by Equation 78. To evaluate the corresponding minus quantities $\mathrm{Tr}(\mathbf{V}_-^n)$ and $\mathrm{Tr}(\mathbf{U}\mathbf{V}_-^n)$ in Equation 57 we use the representation 64 with $\boldsymbol{\sigma}^a\mathbf{r}^-$ in place of $\boldsymbol{\sigma}^a\mathbf{r}^+$ (to preserve the orthogonality of the two terms in Equation 56). The above calculation then goes through with essentially a minus sign in place of a plus (although some special care must be taken with the $q = 0$ and $q = \pi$ terms arising from the cyclic boundary conditions).

If we now define γ_k to be the positive solution of Equation 78 and note that $\gamma_{2m-k} = \gamma_k$ for $k = 0, 1, \ldots, m$ we have after combining Equations 56, 57, 81, and 83 and the corresponding minus results, that

$$\begin{aligned}Z_{n,m} &= (2\sinh 2v)^{mn/2}\,\mathrm{Tr}(\mathbf{V}^n) \\ &= \tfrac{1}{2}(2\sinh 2v)^{mn/2}\left[\prod_{j=1}^{m} 2\cosh\left(\frac{n\gamma_{2j-1}}{2}\right) + \prod_{j=1}^{m} 2\sinh\left(\frac{n\gamma_{2j-1}}{2}\right)\right. \\ &\qquad \left. + \prod_{j=1}^{m} 2\cosh\left(\frac{n\gamma_{2j}}{2}\right) + \prod_{j=1}^{m} 2\sinh\left(\frac{n\gamma_{2j}}{2}\right)\right].\end{aligned} \tag{84}$$

This result was obtained by Kaufman (1949a) using a slightly different method. Kaufman actually found all the eigenvalues of **V** to obtain Equation 84. This can be done also in the present formalism and the interested reader is referred to Schultz et al. (1964) for details. The results are as follows. Corresponding to $\mathbf{V}_+$ and $\mathbf{V}_-$ we have two sets of eigenvalues λ_+ and λ_-, respectively, given by

$$\log \lambda_+ = \tfrac{1}{2}(\pm\gamma_1 \pm \gamma_3 \pm \cdots \pm \gamma_{2m-1})$$

and (85)

$$\log \lambda_- = \tfrac{1}{2}(\pm\gamma_2 \pm \gamma_4 \pm \cdots \pm \gamma_{2m}),$$

where γ_k is defined by Equation 78. The subset of eigenvalues of **V** is obtained from Equation 85 by restricting the number of minus signs to be even in both cases. The two maximal values from Equation 85 are given by (with γ_0 instead of γ_{2m})

$$\lambda_{\max}^+ = \exp[\tfrac{1}{2}(\gamma_1 + \gamma_3 + \cdots + \gamma_{2m-1})]$$

and (86)

$$\lambda_{\max}^- = \exp[\tfrac{1}{2}(\gamma_0 + \gamma_2 + \cdots + \gamma_{2m-2})].$$

For large m it is obvious from the definition Equation (78) that $\gamma_{2r} \sim \gamma_{2r-1} \cdot \gamma_0$ is, however, an exception, and it is this peculiar mathematical fact that is essentially responsible for the phase transition to long-range order. From the detailed analysis of the $\mathbf{V}_-$ terms (which is not given above) the proper definition of γ_0 in Equations 86 and 84 is

$$\gamma_0 = 2(\nu - \nu^*) \tag{87}$$

[compare with Equation 78, which gives $\cosh \gamma_0 = \cosh 2(\nu - \nu^*)$], which *changes sign* at the critical point $\nu = \nu^*$, i.e., from Equation 8 at the point ν_c given by

$$\sinh 2\nu_c = 1. \tag{88}$$

All the other γ_r's are more or less symmetrical about $\nu = \nu_c$. For large m, then, from Equations 78 and 87, $\gamma_1 \sim \gamma_0$ for $\nu > \nu_c$ (i.e., for temperatures below the critical temperature) and $\gamma_1 \sim -\gamma_0$ for $\nu < \nu_c$ (i.e., for temperatures above the critical temperature). It follows from Equations 86 and 87 that

$$\frac{\lambda_{\max}^-}{\lambda_{\max}^+} \sim \begin{cases} 1 - O(e^{-cm}) & \text{for } T < T_c \text{ as } m \to \infty \\ e^{2(\nu-\nu^*)} & \text{for } T > T_c \text{ as } m \to \infty; \end{cases} \tag{89}$$

i.e., the maximum eigenvalue of the transfer matrix is asymptotically degenerate below T_c and strictly nondegenerate above T_c. This phenomenon is related to the existence of long-range order discussed in Section 5-6.

Returning to the partition function Equation 3, we have from Equation 86 that

$$\lim_{m\to\infty}\lim_{n\to\infty}(mn)^{-1}\log Z_{n,m} = \tfrac{1}{2}\log(2\sinh 2v) + \lim_{m\to\infty} m^{-1}\log\lambda_{\max}$$

$$= \tfrac{1}{2}\log(2\sinh 2v) + \lim_{m\to\infty}(2m)^{-1}\sum_{j=1}^{m}\gamma_{2j-1}. \quad (90)$$

This result also follows from Equation 84, since for large m and n,

$$Z_{n,m} \sim (2\sinh 2v)^{mn/2}\exp\left(\frac{n}{2}\sum_{j=1}^{m}\gamma_{2j-1}\right)\left[1+\prod_{j=1}^{m}\tanh\left(\frac{n\gamma_{2j-1}}{2}\right)\right], \quad (91)$$

where we have made use of the fact that for large x, $2\cosh x \sim e^x$. It is obvious that Equation 91 also gives Equation 90, since $|\tanh x| \leq 1$.

This completes our discussion of the algebraic derivation of the two-dimensional Ising-model partition function.

APPENDIX E

Combinatorial Solution of the Dimer and Ising Problems in Two Dimensions

We shall consider here dimer problems on a square lattice and on Fisher's lattice, the latter giving the partition function for the two-dimensional square-lattice Ising model. The presentation given here is based on Montroll's review [Montroll (1968)] of the Pfaffian method.

Our strategy will be to express the dimer-generating functions as Pfaffians, so first it is appropriate to define a Pfaffian.

Consider a $2n$ by $2n$ antisymmetric matrix A with elements $a_{i,j} = -a_{j,i}$. The Pfaffian of A, Pf A, is roughly half a determinant and is usually written in the following way:

$$\text{Pf}\, A = \begin{vmatrix} a_{1,2} & a_{1,3} & \cdots & a_{1,2n} \\ & a_{2,3} & \cdots & a_{2,2n} \\ & & & \vdots \\ & & & a_{2n-1,2n} \end{vmatrix}. \tag{1}$$

The precise definition of Pf A is

$$\text{Pf}\, A = \sum_P \delta_P \, a_{p_1,p_2} a_{p_3,p_4} \cdots a_{p_{2n-1},p_{2n}}, \tag{2}$$

where the sum is over permutations $(p_1, p_2, \ldots, p_{2n})$ of the integers $(1, 2, \ldots, 2n)$ with the restrictions that

$$p_1 < p_2, p_3 < p_4, \ldots, p_{2n-1} < p_{2n}$$

and (3)

$$p_1 < p_3 < \cdots < p_{2n-1}.$$

δ_P in Equation 2 is $+1$ if the permutation P is even, and -1 if P is odd. For example, when $n = 2$ it is easily verified that

$$\begin{vmatrix} a_{1,2} & a_{1,3} & a_{1,4} \\ & a_{2,3} & a_{2,4} \\ & & a_{3,4} \end{vmatrix} = a_{1,2}a_{3,4} - a_{1,3}a_{2,4} + a_{1,4}a_{2,3}. \tag{4}$$

Consider now a typical dimer configuration on a square lattice, for simplicity on a 4 by 4 lattice, shown in Figure E.1. If the lattice points are numbered as shown in Figure E.1, we can represent the dimer configuration by the sequence of pairs of numbers

$$(1, 2) \quad (3, 7) \quad (4, 8) \quad (5, 6) \quad (9, 13) \quad (10, 11) \quad (12, 16) \quad (14, 15),$$

or in general by

$$(p_1, p_2)(p_3, p_4)(p_5, p_6) \cdots (p_{2n-1}, p_{2n}),$$

where $(p_1, p_2, \ldots, p_{2n})$ is a permutation of $(1, 2, \ldots, 2n)$ satisfying the restrictions (3) for permutations appearing in a Pfaffian. If we then define the elements $a_{p,p'}$ in the Pfaffian by

$$\begin{aligned} |a_{p,p'}| &= z_1 && \text{if } p' > p \text{ and } p, p' \text{ are horizontal nearest neighbors} \\ &= z_2 && \text{if } p' > p \text{ and } p, p' \text{ are vertical nearest neighbors} \\ &= 0 && \text{otherwise,} \end{aligned} \tag{5}$$

it is obvious that we have set up a one-to-one correspondence between dimer

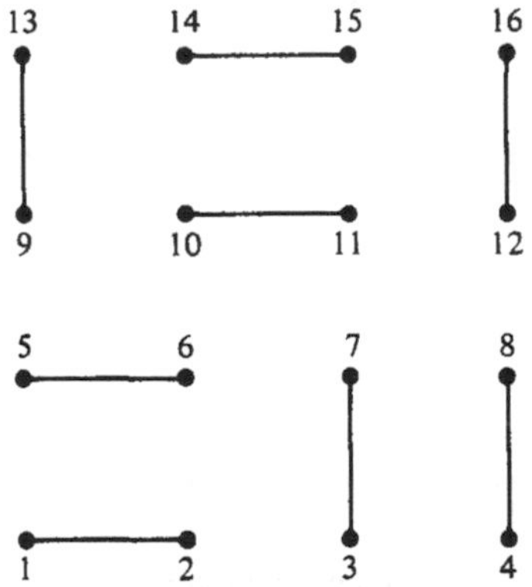

FIGURE E.1. **Dimer configuration on a 4 by 4 square lattice.**

configurations and the terms in the expansion for Pf A. We would naturally like to have

$$\Phi(z_1, z_2) = \text{Pf } A,$$

where (6)

$$\Phi(z_1, z_2) = \sum_{n_1, n_2} g(n_1, n_2) z_1^{n_1} z_2^{n_2}$$

is the dimer-generating function, i.e., $g(n_1, n_2)$ is the number of dimer configurations with n_1 horizontal dimers and n_2 vertical dimers. Equation 6, from the definition of Pf A (Equation 2), means that in addition to Equation 5 we must assign phase factors (or signs) to the elements $a_{p,p'}$, so that all terms in Pf A are positive. It turns out that this can only be done if the lattice is planar, i.e., two-dimensional without crossed bonds or bridges (next-nearest-neighbor interactions therefore, are out). The problem of evaluating Pf A is elementary in view of the classical result

$$\text{Pf } A = (\text{Det } A)^{1/2}. \tag{7}$$

(Recall that A is antisymmetric.)

The only problem remaining then is the sign problem. The first step is to introduce the standard dimer configuration, which consists solely of horizontal dimers, shown in Figure E.2 for a 4 by 4 lattice. We then superimpose the standard configuration on a particular configuration. In Figure E.3 we have shown this superposition for the configuration shown in Figure E.1.

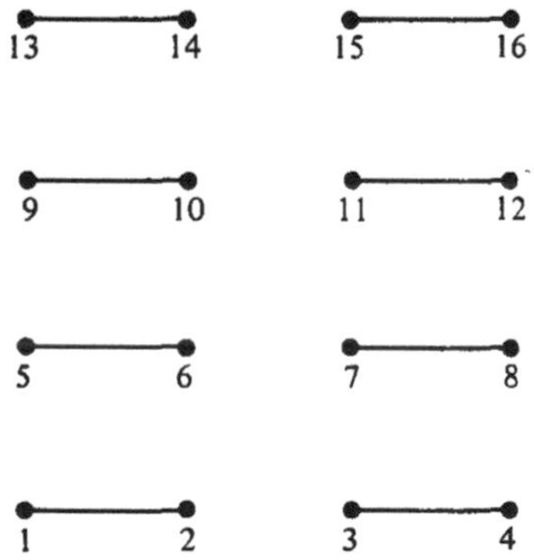

FIGURE E.2. Standard dimer configuration on a 4 by 4 square lattice.

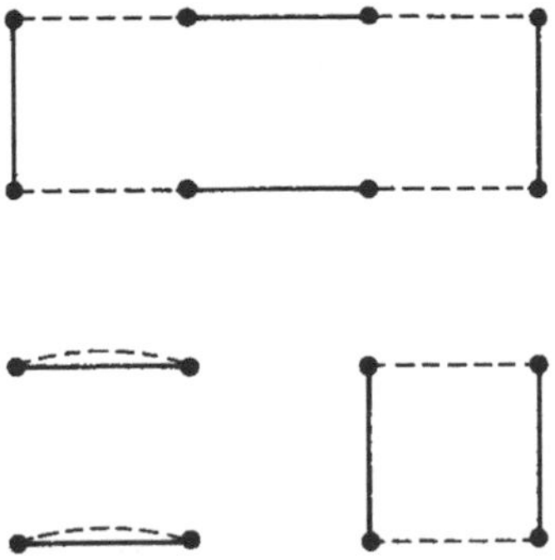

FIGURE E.3. Superposition of the dimer configuration (Figure E.1) and the standard configuration (Figure E.2).

The dashed lines denote the standard dimer configuration. The resulting diagram consists of closed "superposition polygons" and "overlapping dimer pairs."

There are now two points to notice. First, a clockwise displacement of a particular superposition polygon through one lattice spacing transforms the polygon into itself with dashed lines replaced by solid lines and vice versa. Second, the restriction (3) on the terms in the Pfaffian can be violated in particular ways without changing the value of the Pfaffian. For example, if we interchange p_1 and p_2 the inequality $p_1 < p_2$ is violated, but since $a_{p_1,p_2} = -a_{p_2,p_1}$ and δ_P changes sign under this interchange, the value of Pf A (Equation 2) is unchanged. Similarly, if the pairs (p_1, p_2) and (p_3, p_4) are interchanged, δ_P is unchanged and hence Pf A is unchanged. This fact and the rotational invariance of the superposition polygons suggests that the original association of dimer configurations and Pfaffian terms is not the most convenient. A more useful association is obtained by looking at the individual superposition polygons separately and assigning p_j's to the vertices in, say, clockwise order. The particular ordering of the individual polygons is unimportant. Now let P_0 be the (identity) permutation corresponding to the standard configuration and P_1 the permutation corresponding to an arbitrary configuration. From the above remarks it follows that

$$P_1 = P_2 P_0, \tag{8}$$

where P_2 is the product of the cyclic permutations which transform the standard configuration into the arbitrary configuration. In other words, if the arbitrary configuration is composed of s superposition polygons, P_2 consists of s cycles $c_1, c_2, \ldots, c_s$, one for each polygon. Since the polygons must have an even number of sides, each cycle c_i is an odd permutation; i.e., $\delta_{c_i} = -1$ and hence

$$\begin{aligned}\delta_{P_2} &= \delta_{c_1}\delta_{c_2}\cdots\delta_{c_s}\\ &= (-1)^s.\end{aligned} \tag{9}$$

Now if we can make our selection of signs so that each term in the expansion of the Pfaffian has the same sign as that of the standard configuration, every term in the expansion of the Pfaffian would be positive as required. This will certainly be the case if we make the sign of the term associated with the standard configuration, say

$$a_{p_1,p_2}a_{p_3,p_4}\cdots a_{p_{2k-1},p_{2k}},$$

the same as the sign of the term associated with the configuration obtained by a clockwise rotation c, i.e.,

$$\delta_c a_{p_2,p_3}a_{p_4,p_5}\cdots a_{p_{2k},p_1}.$$

Since $\delta c = -1$, as noted above, this means that

$$\operatorname{sgn}(a_{p_1,p_2}a_{p_3,p_4}\cdots a_{p_{2k-1},p_{2k}}) = -\operatorname{sgn}(a_{p_2,p_3}a_{p_4,p_5}\cdots a_{p_{2k},p_1}), \tag{10}$$

i.e., that

$$\prod_{i=1}^{2k}\operatorname{sgn}(a_{p_i,p_{i+1}}) = -1 \qquad (p_{2k+1} \equiv p_1), \tag{11}$$

where the product is over all matrix elements of the superposition polygon associated with the cycle c.

A necessary condition for the validity of Equation 11 is that an odd number of (-1)'s be encountered as one traverses any superposition polygon in a clockwise direction. For the square lattice this condition is fulfilled with the sign prescription shown in Figure E.4. In other words, from Equation 5, taking $p' > p$,

$$a_{p,p'} = \begin{cases} z_1 & \text{for horizontal bonds}\\ (-1)^p z_2 & \text{for vertical bonds}\\ 0 & \text{otherwise.}\end{cases} \tag{12}$$

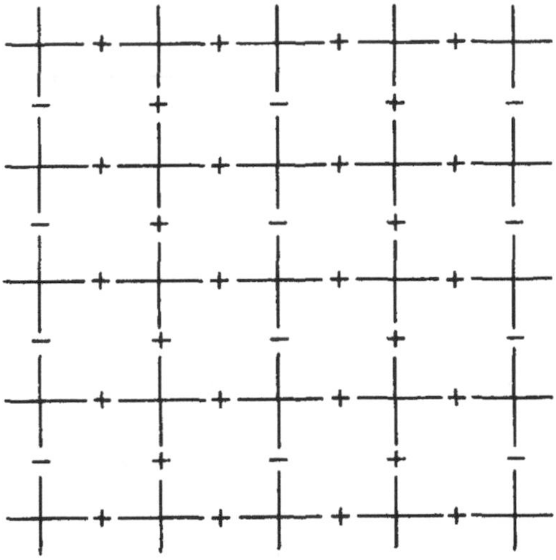

FIGURE E.4. Assignment of signs for the square-lattice dimer problem.

A less cumbersome way of picturing the + and − signs is to associate an arrow with every nearest-neighbor bond on the lattice. Thus, as shown in Figure E.5, an arrow points from p to p' if $a_{p,p'} > 0$ and from p' to p if $a_{p,p'} < 0$. We say that a superposition polygon is clockwise even if the polygon has an even number of clockwise arrows, and clockwise odd if the polygon has an odd number of clockwise arrows. The general problem is to orient the bonds (with arrows) so that all superposition polygons are clockwise odd.

It is not difficult to prove that the orientation shown in Figure E.5 fulfills condition 11 for the square lattice. The prescription leading to this result is, however, a general one and rests on a theorem of Kasteleyn concerning planar graphs. So rather than prove the special case of Equation 12 we will state and prove the general theorem. We need first a few definitions.

1. ***A planar graph*** is a set of points in a plane with bonds connecting the points.

2. ***An elementary polygon*** is a closed planar graph that contains no interior points (e.g., a square for the square lattice).

3. ***An articulation point*** is a point of a graph such that if it is removed the resulting graph consists of two or more disconnected parts. Examples of these three objects are given in Figure E.6. Obviously, a closed planar graph

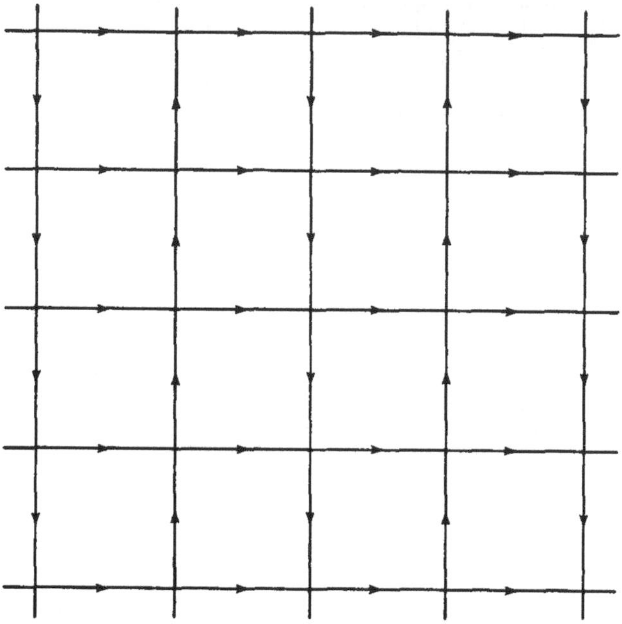

FIGURE E.5. Assignment of arrows for the square-lattice dimer problem. Arrows up and to the right correspond to plus bonds and arrows down correspond to minus bonds in Figure E.4.

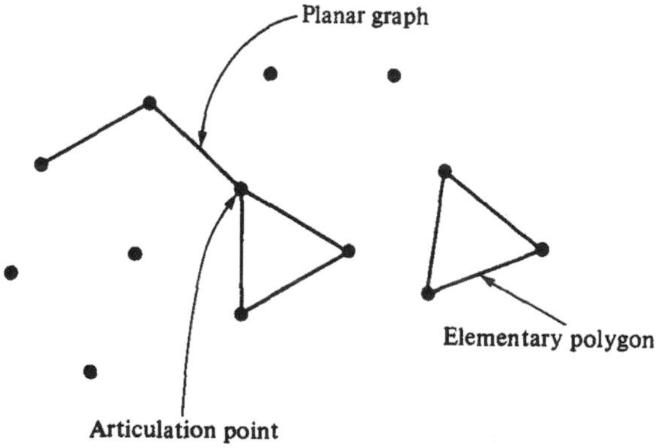

FIGURE E.6. Illustration of some elementary concepts in graph theory.

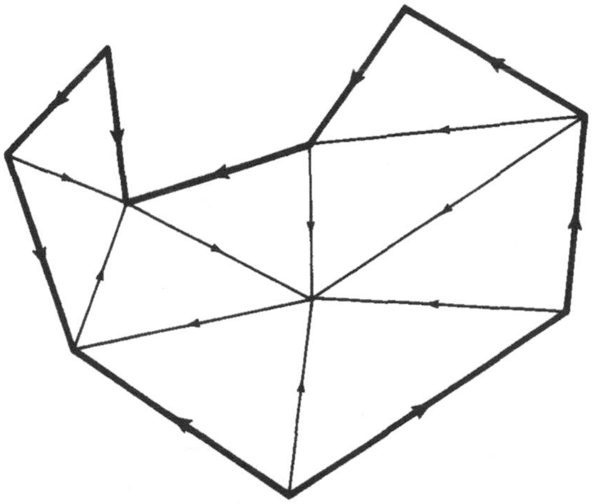

FIGURE E.7. Oriented planar graph decomposed into its clockwise odd-component polygons. The parity of the graph is (clockwise) even and the number of enclosed points (one) is odd.

without crossed bonds can be built out of elementary polygons, with the boundary forming the planar graph as shown in Figure E.7. In this context we call the elementary polygons component polygons.

We now orient the bonds of a closed planar graph without crossed bonds or articulation points as follows. Choose any component polygon and orient the bonds so that it is clockwise odd. Consider now an adjacent component polygon which has at least one bond in common with the first component polygon. Orient the bonds of this polygon so that it is also clockwise odd. Clearly this can always be done since each adjacent component polygon has at least one bond which is not oriented. Continuing in this way we obtain an orientation for the bonds on the planar graph. An example is given in Figure E.7 and orientations for the triangular and Fisher lattices are given in Figure E.8.

Now we state

KASTELEYN'S THEOREM. *Consider a closed oriented planar graph without articulation points or crossed bonds such that the component polygons are all clockwise odd. The parity of the clockwise bonds of such a graph is then opposite to the parity of the number of enclosed points, i.e., the graph is clockwise even (odd) if the number of enclosed points is odd (even).*

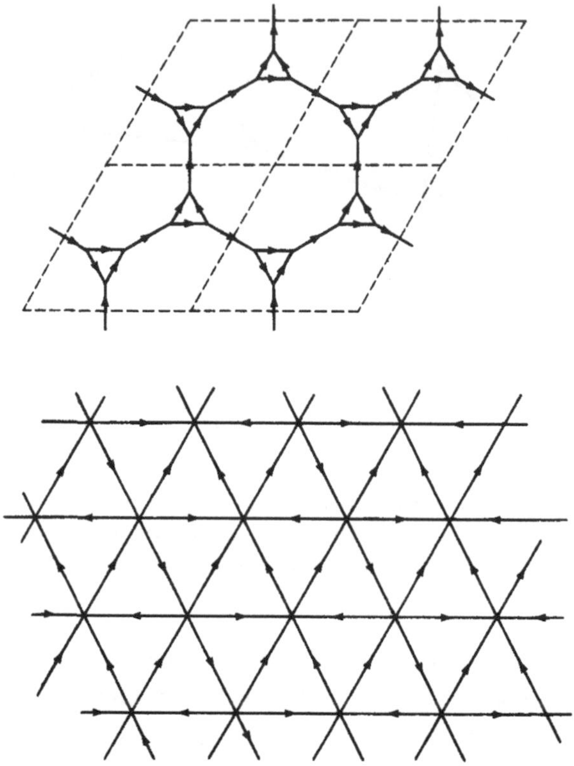

FIGURE E.8. Assignment of arrows on Fisher's lattice and on the triangular lattice.

An example of this theorem is Figure E.7: The graph is clockwise even and the number of enclosed points is one.

As a

COROLLARY. *Superposition polygons are clockwise odd if all elementary polygons are clockwise odd.*

This is the result we need to get all positive signs in the expansion of the dimer Pfaffian.

To prove the corollary we have simply to show, in view of the theorem, that a superposition polygon must enclose an even number of points.

Recall that superposition polygons are obtained by superimposing the standard dimer configuration on an arbitrary dimer configuration. It follows that a superposition polygon contains the same number of bonds from the standard configuration as from the arbitrary configuration, and hence that the number of points (and bonds) making up a superposition polygon is even. Now since in the superposition all lattice points are part of a superposition polygon or a dimer pair, a given superposition polygon contains either no interior points, or interior points that are part of other superposition polygons or dimer pairs. Superposition polygons therefore enclose an even number of points, and the corollary is proved.

We now prove the theorem by induction.

By construction, all elementary polygons satisfy the theorem, since they are clockwise odd and contain no interior points. Let Γ_n be a polygon containing n elementary polygons Γ_1. Assuming the validity of the theorem for Γ_n, we wish to prove the validity of the theorem for Γ_{n+1} obtained by adjoining a Γ_1 to a Γ_n. Let α be the number of clockwise arrows on Γ_n and let γ be the number of interior points of Γ_n. By hypothesis, α and γ have opposite parity. Now if β is the number of bonds common to Γ_n and Γ_1, Γ_{n+1} obviously contains $\gamma + \beta - 1$ interior points. Similarly, the number of clockwise arrows on Γ_{n+1} is $\alpha + \alpha' - \beta$, where α' is the number of clockwise arrows on Γ_1, which is odd by definition. This is because the β common bonds must be anticlockwise on Γ_n or on Γ_1. The parity of Γ_{n+1} is therefore even (odd) if $\alpha + \alpha' - \beta$ is even (odd). Now since α and γ have opposite parity and α' is odd, $\alpha + \alpha'$ and $\gamma - 1$ have opposite parity. It follows that $\alpha + \alpha' - \beta$ has opposite parity to the number of interior points $\gamma - 1 + \beta$ and the theorem is proved.

Let us now return to the square-lattice problem. The first thing to note is that the pattern of arrows in Figure E.5 is doubly periodic in the horizontal direction. It is, therefore, appropriate to consider the oriented lattice as made up of unit cells containing two nearest-neighbor horizontal lattice points as shown in Figure E.9. It is then convenient to represent each unit cell by (p_1, p_2), where p_1 denotes the horizontal position of the cell and p_2 the vertical position of the cell. The two lattice points in the (p_1, p_2) cell are then denoted by (p_1, p_2, L) and (p_1, p_2, R), as shown in Figure E.9. From Equation 12 the only nonzero matrix elements of A are

$$\begin{aligned} a(p_1, p_2, L; p_1, p_2, R) &= z_1, \\ a(p_1, p_2, R; p_1 + 1, p_2, L) &= z_1, \\ a(p_1, p_2, R; p_1, p_2 + 1, R) &= z_2, \\ a(p_1, p_2, L; p_1, p_2 + 1, L) &= -z_2. \end{aligned} \tag{13}$$

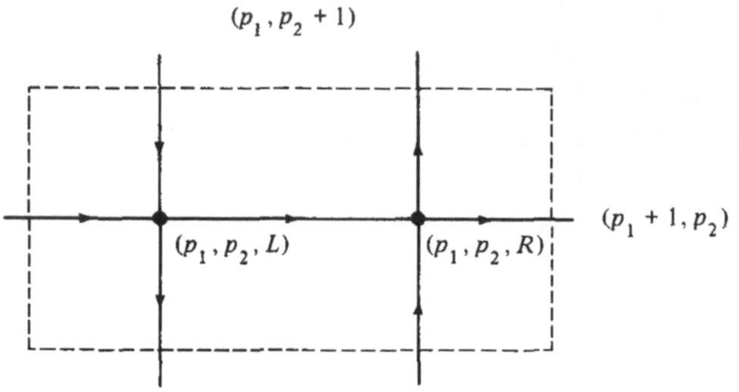

FIGURE E.9. Unit cell (p_1, p_2) for the oriented square lattice.

To express the matrix A in a convenient form we define the 2 by 2 matrix $A(p_1, p_2; p_1', p_2')$ by

$$A(p_1, p_2; p_1', p_2') = \begin{pmatrix} a(p_1, p_2, L; p_1', p_2', L) & a(p_1, p_2, L; p_1', p_2', R) \\ a(p_1, p_2, R; p_1', p_2', L) & a(p_1, p_2, R; p_1', p_2', R) \end{pmatrix}. \tag{14}$$

From Equation 13 the only nonzero $A(p_1, p_2; p_1', p_2')$ matrices are

$$\begin{aligned} A(p_1, p_2; p_1, p_2) &= \begin{pmatrix} 0 & z_1 \\ -z_1 & 0 \end{pmatrix} = \alpha(0, 0), \\ A(p_1, p_2; p_1+1, p_2) &= \begin{pmatrix} 0 & 0 \\ z_1 & 0 \end{pmatrix} = \alpha(1, 0), \\ A(p_1, p_2; p_1-1, p_2) &= \begin{pmatrix} 0 & -z_1 \\ 0 & 0 \end{pmatrix} = \alpha(-1, 0), \\ A(p_1, p_2; p_1, p_2+1) &= \begin{pmatrix} -z_2 & 0 \\ 0 & z_2 \end{pmatrix} = \alpha(0, 1), \end{aligned} \tag{15}$$

and

$$A(p_1, p_2; p_1, p_2-1) = \begin{pmatrix} z_2 & 0 \\ 0 & -z_2 \end{pmatrix} = \alpha(0, -1).$$

The antisymmetric matrix A whose Pfaffian gives the generating function for the dimer problem on a square lattice with n rows and $2m$ columns (so that the number of horizontal unit cells is m) is then given by

$$
\begin{array}{c}
\\ (1,1) \\ (2,1) \\ \\ \\ (m,1) \\ (1,2) \\ \\ \\ (m,2) \\ \\ \\ (m,n)
\end{array}
\begin{array}{cccccc|ccccc|cccc}
(1,1) & (2,1) & (3,1) & & & (m,1) & (1,2) & & & & (m,2) & & & & (m,n) \\
\alpha(0,0) & \alpha(1,0) & 0 & 0 & \cdots & 0 & \alpha(0,1) & 0 & & \cdots & 0 & 0 & 0 & \cdots & 0 \\
\alpha(-1,0) & \alpha(0,0) & \alpha(1,0) & 0 & \cdots & 0 & 0 & \alpha(0,1) & & & & 0 & 0 & & 0 \\
0 & \alpha(-1,0) & \alpha(0,0) & \alpha(1,0) & \cdots & 0 & 0 & & \ddots & & & & \vdots & & \\
\vdots & \vdots & \vdots & \vdots & \ddots & \vdots & \vdots & & & & & & & & \\
0 & 0 & 0 & 0 & & \alpha(0,0) & 0 & & & & \alpha(0,1) & 0 & 0 & \cdots & 0 \\
\hline
\alpha(0,-1) & 0 & 0 & & & 0 & \alpha(0,0) & \alpha(1,0) & 0 & \cdots & 0 & \alpha(0,1) & 0 & \cdots & 0 \\
0 & \alpha(0,-1) & 0 & & & & \alpha(-1,0) & \alpha(0,0) & \alpha(1,0) & \cdots & 0 & 0 & \alpha(0,1) & & \\
 & & \ddots & & & & 0 & \alpha(-1,0) & & \cdots & 0 & & & & \\
0 & & & & & \alpha(0,-1) & 0 & & & & \alpha(0,0) & & & & \\
\hline
0 & & & & & & & & & & & & & & \\
\vdots & & & & & & & & & & & & & & \\
0 & & & & & 0 & & & & & & & & & 0
\end{array}
\qquad (16)
$$

The evaluation of the determinant of A (and hence from Equation 7 the Pfaffian of A) is made somewhat easier by imposing periodic boundary conditions, i.e., by taking

$$\alpha(k+m, l+n) = \alpha(k, l). \tag{17}$$

In view of the translational invariance of the matrix, i.e., from Equation 15,

$$A(p_1, p_2; p_1', p_2') = \alpha(p_1' - p_1, p_2' - p_2), \tag{18}$$

the condition 17 is equivalent to wrapping the $2m$ by n lattice on a torus. Equations 17 and 18 mean that the matrix A is cyclic.

In general we say that a matrix A with elements $A(\mathbf{p}, \mathbf{q})$ (which may themselves be matrices), and $\mathbf{p} = (p_1, \ldots, p_s)$, $\mathbf{q} = (q_1, \ldots, q_s)$ s-dimensional vectors, is cyclic if

$$A(\mathbf{p}, \mathbf{q}) = A(\mathbf{p} - \mathbf{q})$$

and (19)

$$A(\mathbf{p} + \mathbf{n}) = A(\mathbf{p}),$$

where $\mathbf{n}$ is a fixed s-dimensional vector. Our case, Equations 16–18, corresponds to $s = 2$, but let us consider first the case $s = 1$. From Equation 19, with $\mathbf{n} = n$, the matrix A can be written as

$$A = \begin{pmatrix} A(0) & A(1) & A(2) & \cdots & A(n-1) \\ A(n-1) & A(0) & A(1) & \cdots & A(n-2) \\ A(n-2) & A(n-1) & A(0) & \cdots & A(n-3) \\ \vdots & & & & \vdots \\ A(1) & A(2) & A(3) & & A(0) \end{pmatrix}. \tag{20}$$

The matrix S which diagonalized this matrix has components

$$S_{k,l} = n^{-1/2} \exp\left(\frac{2\pi ikl}{n}\right). \tag{21}$$

In view of the orthogonality property of the Fourier coefficients,

$$(S^{-1})_{k,l} = n^{-1/2} \exp\left(-\frac{2\pi ikl}{n}\right), \tag{22}$$

and hence

$$
\begin{aligned}
(S^{-1}AS)_{k,l} &= \sum_{s,t=1}^{n} S_{k,s}^{-1} A_{s,t} S_{t,l} \\
&= n^{-1} \sum_{s=1}^{n} \exp\left[\frac{2\pi i(l-k)s}{n}\right] \sum_{t=1}^{n} A(t-s) \exp\left[\frac{2\pi i(t-s)l}{n}\right] \\
&= n^{-1} \sum_{s=1}^{n} \exp\left[\frac{2\pi i(l-k)s}{n}\right] \sum_{u=1}^{n} A(u) \exp\left(\frac{2\pi i u l}{n}\right) \qquad (23) \\
&= \delta_{k,l}\, \lambda\left(\frac{2\pi k}{n}\right),
\end{aligned}
$$

where

$$\lambda(\phi) = \sum_{s=1}^{n} A(s) \exp(is\phi), \tag{24}$$

and in the last step, Equation 23, we have used the orthogonality property of the Fourier coefficients and the cyclic property of the matrix elements, Equation 19.

For the general s-dimensional case, Equation 19, the diagonalizing matrix is

$$S_{\mathbf{k},\mathbf{l}} = \prod_{\alpha=1}^{s} n_\alpha^{-1/2} \exp\left(\frac{2\pi i k_\alpha l_\alpha}{n_\alpha}\right) \tag{25}$$

and

$$(S^{-1}AS)_{\mathbf{k},\mathbf{l}} = \delta_{\mathbf{k},\mathbf{l}}\, \lambda\left(\frac{2\pi k_1}{n_1}, \ldots, \frac{2\pi k_s}{n_s}\right), \tag{26}$$

where

$$\lambda(\phi_1, \phi_2, \ldots, \phi_s) = \sum_{\mathbf{u}} A(\mathbf{u}) \exp(i\mathbf{u} \cdot \boldsymbol{\phi}). \tag{27}$$

The results, Equations 23 and 26, are also valid when the elements $A(\mathbf{u})$ are themselves matrices. In general, then, the matrix consists of blocks of matrices down the main diagonal, so the determinant of A is the product of the determinants of the λ's. That is, from Equation 26,

$$\left(\prod_{\alpha=1}^{s} n_\alpha^{-1}\right) \log \operatorname{Det}(A) = \left(\prod_{\alpha=1}^{s} n_\alpha^{-1}\right) \sum_{k_1=1}^{n_1} \cdots \sum_{k_s=1}^{n_s} \log \operatorname{Det} \lambda\left(\frac{2\pi k_1}{n_1}, \ldots, \frac{2\pi k_s}{n_s}\right). \tag{28}$$

In the limit $n_\alpha \to \infty$, $\alpha = 1, 2, \ldots, s$, the right-hand side of Equation 28 becomes an integral; i.e., with $\phi_\alpha = 2\pi k_\alpha / n_\alpha$,

$$\lim_{\substack{n_\alpha \to \infty \\ \alpha = 1, 2, \ldots, s}} \left(\prod_{\alpha=1}^{s} n_\alpha^{-1} \right) \log \operatorname{Det}(A) = (2\pi)^{-s} \int_0^{2\pi} \cdots \int \log \operatorname{Det} \lambda(\phi_1, \ldots, \phi_s)\, d\phi_1 \cdots d\phi_s. \tag{29}$$

For the problem at hand, Equations 16–18,

$$\lim_{m, n \to \infty} (mn)^{-1} \log \operatorname{Det}(A) = (2\pi)^{-2} \iint_0^{2\pi} \log[\operatorname{Det} \lambda(\phi_1, \phi_2)]\, d\phi_1\, d\phi_2, \tag{30}$$

where, from Equations 15 and 27,

$$\begin{aligned} \lambda(\phi_1, \phi_2) &= \sum_{p_1, p_2} \alpha(p_1, p_2) \exp(ip_1\phi_1 + ip_2\phi_2) \\ &= \alpha(0, 0) + \alpha(1, 0)e^{i\phi_1} + \alpha(-1, 0)e^{-i\phi_1} \\ &\quad + \alpha(0, 1)e^{i\phi_2} + \alpha(0, -1)e^{-i\phi_2} \\ &= \begin{pmatrix} z_2 e^{-i\phi_2} - z_2 e^{i\phi_2} & z_1 - z_1 e^{-i\phi_1} \\ z_1 e^{i\phi_1} - z_1 & z_2 e^{i\phi_2} - z_2 e^{-i\phi_2} \end{pmatrix}. \end{aligned} \tag{31}$$

It follows from Equations 30 and 31 and the fact that $\operatorname{Pf} A = (\operatorname{Det} A)^{1/2}$ that

$$\begin{aligned} \lim_{m, n \to \infty} 2(mn)^{-1} \log \operatorname{Pf}(A) &= (2\pi)^{-2} \iint_0^{2\pi} \log 4(z_1^2 \sin^2 \tfrac{1}{2}\phi_1 + z_2^2 \sin^2 \tfrac{1}{2}\phi_2)\, d\phi_1\, d\phi_2 \\ &= (2\pi)^{-2} \iint_0^{2\pi} \log 2[(z_1^2 + z_2^2) - z_1^2 \cos \phi_1 - z_2^2 \cos \phi_2]\, d\phi_1\, d\phi_2. \end{aligned} \tag{32}$$

The generating function $\Phi(z_1, z_2)$ for the dimer problem in the limit of an infinite lattice is then given by Equation 32, since, from Equation 6, $\Phi(z_1, z_2) = \operatorname{Pf} A$. The similarity of Equation 32 and the free energy for the two-dimensional Ising model, Equation 3.7 of Chapter 6, is to be noted.

To solve the dimer problem we have to evaluate Equation 32 when $z_1 = z_2 = 1$. This leads straightforwardly to the result

$$\lim_{m, n \to \infty} 2(mn)^{-1} \log \Phi(1, 1) = \frac{4G}{\pi}, \tag{33}$$

where

$$G = 1 - 3^{-2} + 5^{-2} - 7^{-2} + \cdots = 0.9159655\ldots \tag{34}$$

is *Catalan's constant.* Recall that $\Phi(1, 1)$ is the number of dimer coverings of the lattice.

Exact results for finite lattices can be obtained by the above method, but this involves considerably more manual labor. An interesting result, due to Fisher, in this connection is the number of ways of covering a chess board (8 by 8) with 32 dominos, each domino covering two squares. The result is $2^4(901)^2$!

Let us now turn to the dimer problem on Fisher's lattice. The solution of this problem, as was shown in Section 6-3, gives the partition function for the two-dimensional Ising model.

Consider the unit cell of the Fisher lattice (Figure 6.10) with arrows and weights as shown in Figure E.10 (compare with Figure E.8). For the Ising problem with horizontal interaction constant J_1 and vertical interaction constant J_2 we choose

$$\omega_3 = z_1 = z_2 = z_3 = z_4 = z_5 = z_6 = 1$$

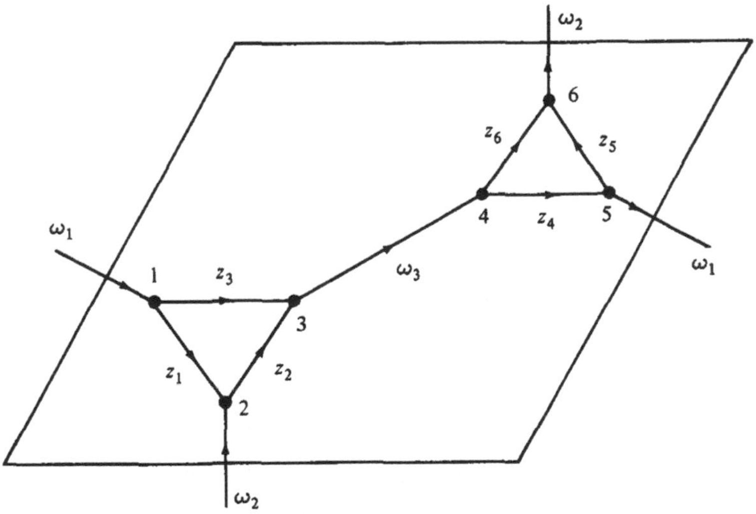

FIGURE E.10. Assignment of weights for the dimer problem on Fisher's lattice.

and

$$\omega_1 = \tanh\left(\frac{J_1}{kT}\right), \qquad \omega_2 = \tanh\left(\frac{J_2}{kT}\right). \tag{35}$$

The partition function is then given by Equation 3.4 of Chapter 6.

To solve the dimer problem on Fisher's lattice we repeat all the steps in the above argument essentially word for word. The only difference is that the unit cell now contains six lattice points rather than two for the square lattice. This means that the matrices $\alpha(p_1, p_2)$ in Equations 15 are 6 by 6 rather than 2 by 2. As for the square lattice, the only nonzero matrices $\alpha(p_1, p_2)$ are $\alpha(\pm 1, 0)$, $\alpha(0, \pm 1)$, and $\alpha(0, 0)$. The reader may easily verify that

$$\alpha(0,0) = \begin{array}{c|cccccc} \backslash & 1 & 2 & 3 & 4 & 5 & 6 \\ \hline 1 & 0 & z_1 & z_3 & 0 & 0 & 0 \\ 2 & -z_1 & 0 & z_2 & 0 & 0 & 0 \\ 3 & -z_3 & -z_2 & 0 & \omega_3 & 0 & 0 \\ 4 & 0 & 0 & -\omega_3 & 0 & z_4 & z_6 \\ 5 & 0 & 0 & 0 & -z_4 & 0 & z_5 \\ 6 & 0 & 0 & 0 & -z_6 & -z_5 & 0 \end{array}. \tag{36}$$

For example, the 3, 1 element corresponds to the bond connecting lattice points 3 to 1, which has weight $-z_3$, as shown in Figure E.10. Similarly, $\alpha(1, 0)$ has only one nonzero element, ω_1, in the fifth row and first column; $\alpha(-1, 0)$ only one nonzero element, $-\omega_1$, in the first row and fifth column; $\alpha(0, 1)$ only one nonzero element, ω_2, in the sixth row and second column; and $\alpha(0, -1)$ only one nonzero element, $-\omega_2$, in the second row and sixth column. With the α matrices we obtain, from Equations 27 and 31,

$$\lambda(\phi_1, \varphi_2) = \begin{pmatrix} 0 & z_1 & z_3 & 0 & -\omega_1 e^{-i\phi_1} & 0 \\ -z_1 & 0 & z_2 & 0 & 0 & -\omega_2 e^{-i\phi_2} \\ -z_3 & -z_2 & 0 & \omega_3 & 0 & 0 \\ 0 & 0 & -\omega_3 & 0 & z_4 & z_6 \\ \omega_1 e^{i\phi_1} & 0 & 0 & -z_4 & 0 & z_5 \\ 0 & \omega_2 e^{i\phi_2} & 0 & -z_6 & -z_5 & 0 \end{pmatrix}, \tag{37}$$

so

$$\begin{aligned} \text{Det}\,\lambda(\phi_1, \phi_2) = {} & z_1^2 z_5^2 \omega_3^2 + z_3^2 z_4^2 \omega_2^2 + z_2^2 z_6^2 \omega_1^2 + \omega_1^2 \omega_2^2 \omega_3 \\ & + 2\omega_1\omega_2(z_2 z_3 z_4 z_6 - \omega_3^2 z_1 z_5)\cos(\phi_1 - \phi_2) \\ & + 2\omega_2 \omega_3(z_1 z_3 z_4 z_5 - \omega_1^2 z_2 z_6)\cos \phi_2 \\ & + 2\omega_3 \omega_1(z_1 z_2 z_5 z_6 - \omega_2^2 z_3 z_4)\cos \phi_1. \end{aligned} \tag{38}$$

For the Ising problem, Equation 35,

$$\operatorname{Det} \lambda(\phi_1, \phi_2) = (1 + \omega_1^2)(1 + \omega_2^2) \\ - 2\omega_1(1 - \omega_2^2)\cos \phi_1 - 2\omega_2(1 - \omega_1^2)\cos \phi_2 \tag{39}$$

and from Equations 29 and 3.4 of Chapter 6 it follows that the free energy per spin ψ for the Ising problem is given by

$$-\frac{\psi}{kT} = \lim_{N\to\infty} N^{-1} \log Z_N$$

$$= \log 2 + (2\pi)^{-2} \iint_0^{2\pi} \log(\cosh 2v_1 \cosh 2v_2 \tag{40}$$

$$- \sinh v_1 \cos \phi_1 - \sinh v_2 \cos \phi_2)\, d\phi_1\, d\phi_2,$$

where N is the number of lattice sites.

This completes our discussion of the dimer solution of the two-dimensional Ising problem.

Bibliography

ALDER, B. J.

1962. and WAINWRIGHT, T. E., Phase transition in elastic disks, *Phys. Rev.* **127**, 359.

BAKER, G. A., JR.

1961. Applications of the Padé approximant method to the investigation of some magnetic properties of the Ising model, *Phys. Rev.* **124**, 768.

1965. The theory and application of the Padé approximant method, *Advan. Theoret. Phys.* **1**, 1.

1967. and GAUNT, D. S., Ising model critical indices below the critical temperature, *Phys. Rev.* **155**, 545.

BETHE, H. A.

1935. Statistical theory of superlattices, *Proc. Roy. Soc.* (*London*) **A150**, 552.

BOLTZMANN, L.

1868. Studien über das Gleichgewicht der Lebendigen Kraft zwischen bewegten materiellen Punkten, *Wien. Ber.* **58**, 517. (Reprinted in Boltzmann's *Abhandlungen*, Vol. 1, Barth, Leipzig, 1909, p. 49.)

1872. Weitere Studien über das Warmegleichgewicht unter Gasmolekülen, *Wien. Ber.* **66**, 275. (Reprinted in Boltzmann's *Abhandlungen*, Vol. 1, Barth, Leipzig, 1909, p. 316.)

BRAGG, W. L.

1934. and WILLIAMS, E. J., Effect of thermal agitation on atomic arrangement in alloys, I, *Proc. Roy. Soc.* (*London*) **A145**, 699.

1935. and WILLIAMS, E. J., Effect of thermal agitation on atomic arrangement in alloys, II, *Proc. Roy. Soc.* (*London*) **A151**, 540.

BROWN, S. C.

1950. The caloric theory of heat, *Am. J. Phys.* **18**, 319.

BRUSH, S. G.

1967. History of the Lenz–Ising model, *Rev. Mod. Phys.* **39**, 883.

CARATHÉODORY, C.

1909. Untersuchungen über die Grundlagen der Thermodynamik, *Math. Ann.* **67**, 355.

CARNOT, S.

1824. Reflexions sur la puissance motrice du feu et sur les machines propres à développer cette puissance, Bachelier, Paris. English translation in *Reflections on the Motive Power of Fire* . . . , E. Mendoza, ed., Dover, New York, 1960.

CHANGEUX, J. P.

1965. The control of biochemical reactions, *Sci. Am.* **212**, No. 4, 244.

1968a. GERHART, J. C., and SCHACHMAN, H. K., Allosteric interactions in aspartate transcarbamylase, I. Binding of specific ligands to the native enzyme and its isolated subunits, *Biochemistry* **7**, 531.

1968b. and RUBIN, M., Allosteric interactions in aspartate transcarbamylase, III. Interpretation of experimental data in terms of the model of Monod, Wyman and Changeux, *Biochemistry* **7**, 553.

CHARGAFF, E.

1950. Chemical specificity of nucleic acids and mechanism of their enzymatic degradation, *Experientia* **6**, 201.

CLAUSIUS, R.

1850. Über die bewegende Kraft der Wärme und die Gesetze die sich daraus für die Wärmelehre selbst ableiten lassen, *Ann. Phys.* **79**, 368, 500. English translation: *Phil. Mag.* **2**, 1 (1851). Also *Reflections on the Motive Power of Fire* . . . , E. Mendoza, ed., Dover, New York, 1960.

1857. Über die Art der Bewegungen welche wir Wärme Nennen, *Ann. Phys.* **100**, 253. English translation: The nature of the motion which we call heat, *Phil. Mag.* **14**, 108 (1857). Also Brush (B1965), Vol. 1, p. 111.

DALTON, N. W.

1969. and WOOD, D, W., Critical point behavior of the Ising model with higher neighbor interactions present, *J. Math. Phys.* **10**, 1271.

DAVY, H.

1799. Essay on heat, light and the combinations of light, *Contributions to Physical and Medical Knowledge Primarily from the West of England*, collected by T. Beddoes, Bristol, 1799.

DOMB, C.

1949. Order-disorder statistics, II. A two dimensional model, *Proc. Roy. Soc.* (*London*) **A199**, 199.

1957. and SYKES, M. F., On the susceptibility of a ferromagnet above the Curie point, *Proc. Roy. Soc.* (*London*) **A240**, 214.

1960. On the theory of cooperative phenomena in crystals, *Adv. Phys.* **9**, 150.

1961. and Sykes, M. F., Use of series expansions for the Ising model susceptibility and excluded volume problem, *J. Math. Phys.* **2**, 63 [see also *J. Math. Phys.* **3**, 586 (1963)].

Dorfman, J. R.

1967. and Cohen, E. G. D., Difficulties in the kinetic theory of dense gases, *J. Math. Phys.* **8**, 282.

Dresden, M.

1962. A study of models in non-equilibrium statistical mechanics, in *Studies in Statistical Mechanics*, Vol. 1, J. DeBoer, and G. E. Uhlenbeck, eds.) North-Holland, Amsterdam.

Dyson, F. J.

1967. and Lenard, A., Stability of matter, I, *J. Math. Phys.* **8**, 423.

1969a. Existence of a phase-transition in a one dimensional Ising ferromagnet, *Commun. Math. Phys.* **12**, 91.

1969b. Non-existence of spontaneous magnetization in a one dimensional Ising ferromagnet, *Commun. Math. Phys.* **12**, 212.

Edwards, S. F.

1965. The statistical mechanics of polymers with excluded volume, *Proc. Phys. Soc.* **85**, 613.

Essam, J. W.

1963a. and Fisher, M. E., Padé approximant studies of the lattice gas and Ising ferromagnet below the critical point, *J. Chem. Phys.* **38**, 802.

1963b. and Sykes, M. F., The crystal statistics of the diamond lattice, *Physica* **29**, 378.

Fisher, M. E.

1959a. Transformations of Ising models, *Phys. Rev.* **113**, 969.

1959b. The susceptibility of the plane Ising model, *Physica* **25**, 521.

1964. The free energy of a macroscopic system, *Arch. Rat. Mech. Anal.* **17**, 377.

1965. The nature of critical points, in *Lectures in Theoretical Physics*, Vol. VIIC, University of Colorado Press, Boulder, Colo.

1966. On the dimer solution of planar Ising models, *J. Math. Phys.* **7**, 1776.

1967. The theory of equilibrium critical phenomena, *Rept. Progr. Phys.* **30**, 615.

Fowler, R. H.

1937. and Rushbrooke, G. S., Statistical theory of perfect solutions, *Trans. Faraday Soc.* **33**, 1272.

GALLAVOTTI, G.

1968. MIRACLE-SOLE, S., and RUELLE, D., Absence of phase transitions in one dimensional systems with hard cores, *Phys. Letters* **26A**, 350.

GATES, D. J.

1969. and PENROSE, O., The van der Waals limit for classical systems, I. A variational principle, *Commun. Math. Phys.* **15**, 255.

GAUNT, D. S.

1970. and BAKER, G. A., JR., Low temperature critical exponents from high temperature series: The Ising model, *Phys. Rev.* **18**, 1184.

GERHART, J. C.

1963. and PARDEE, A. B., The effect of the feedback inhibitor CTP on subunit interactions in aspartate transcarbamylase, *Cold Spring Harbor Symp. Quant. Biol.* **28**, 491.

1965. and SCHACHMAN, H. K., Distinct subunits for the regulation and catalytic activity of aspartate transcarbamylase, *Biochemistry* **4**, 1054.

1968. and SCHACHMAN, H. K., Allosteric interactions in aspartate transcarbamylase, II. Evidence for different conformation states of the protein in the presence and absence of specific ligands, *Biochemistry* **7**, 583.

GINIBRE, J.

1967. Rigorous lower bound on the compressibility of a classical system, *Phys. Letters* **24A**, 223.

GOEL, N. S.

1968. and MONTROLL, E. W., Denaturation and renaturation of DNA, II. Possible use of synthetic periodic copolymers to establish model parameters, *Biopolymers* **6**, 731.

GRAD, H.

1952. Statistical mechanics, thermodynamics, and fluid dynamics of systems with an arbitrary number of integrals, *Commun. Pure Appl. Math.* **5**, 455.

GRIFFITHS, R. B.

1964. Peierls proof of spontaneous magnetization in a two-dimensional Ising ferromagnet, *Phys. Rev.* **136A**, 437.

1965a. Thermodynamic inequality near the critical point for ferromagnets and fluids, *Phys. Rev. Letters* **14**, 623.

1965b. Microcanonical ensemble in quantum statistical mechanics, *J. Math. Phys.* **6**, 1447.

1966. Spontaneous magnetization of idealized ferromagnets, *Phys. Rev.* **152**, 240.

GUGGENHEIM, E. A.

1945. The principle of corresponding states, *J. Chem. Phys.* **13**, 253.

GUTTMANN, A. J.

1968. NINHAM, B. W., and THOMPSON, C. J., Determination of critical behavior in lattice statistics from series expansions, I, *Phys. Rev.* **172**, 554.

1969a. Determination of critical behavior in lattice statistics from series expansions, III, *J. Phys.* (*C*)**2**, 1900.

1969b. and THOMPSON, C. J., Low temperature susceptibility exponent for the three dimensional Ising model, *Phys. Letters* **28A**, 679.

1970. THOMPSON, C. J., and NINHAM, B. W., Determination of critical behavior in lattice statistics from series expansions, IV, *J. Phys.* (*C*)**3**, 1641.

HAMMERSLEY, J. M.

1957. Percolation processes, II. The connective constant, *Proc. Cambridge Phil. Soc.* **53**, 642.

HELLER, P.

1967. Experimental investigations of critical phenomena, *Rept. Progr. Phys.* **30**, 731.

HEMMER, P. C.

1964. KAC, M., and UHLENBECK, G. E., On the van der Waals theory of the vapor–liquid equilibrium, III. Discussion of the critical region, *J. Math. Phys.* **5**, 60.

HERAPATH, J.

1821. A mathematical inquiry into the causes, laws and principal phenomena of heat, gases, gravitation, etc. *Ann. Philosophy* **1**, 273, 340, 401.

HURST, C. A.

1960. and GREEN, H. S., New solution of the Ising problem for a rectangular lattice, *J. Chem. Phys.* **33**, 1059.

ISING, E.

1925. Beitrag zur Theorie des Ferromagnetismus, *Z. Physik* **31**, 253.

JAYNES, E. T.

1957. Information theory and statistical mechanics, *Phys. Rev.* **106**, 620.

JOULE, J. P.

1845. On the existence of an equivalent relation between heat and the ordinary forms of mechanical power, *Phil. Mag.* **27**, 205.

1848. Some remarks on heat and the constitution of elastic fluids, *Memoirs of the Literary and Philosophical Society of Manchester* **9**, 107 (1851) (paper read in 1848).

KAC, M.

1952. and WARD, J. C., A combinatorial solution of the two dimensional Ising model, *Phys. Rev.* **88**, 1332.

1956. Some remarks on the use of probability in classical statistical mechanics, *Bull. Acad. Roy. Belg.* **42**, 356.

1963. UHLENBECK, G. E., and HEMMER, P. C., On the van der Waals theory of the vapor–liquid equilibrium, I. Discussion of a one-dimensional model, *J. Math. Phys.* **4**, 216.

1968. *Mathematical Mechanisms of Phase Transitions* (Brandeis Lectures 1966), Gordon & Breach, New York.

KADANOFF, L.

1966. Spin-spin correlations in the two-dimensional Ising model, *Nuovo Cimento* **44**, 276.

1967. GÖTZE, W., HAMBLEN, D., HECHT, R., LEWIS, E. A. S., PALCIAUSKAS, V. V., RAYL, M., and SWIFT, J., Static phenomena near critical points: Theory and experiment, *Rev. Mod. Phys.* **39**, 395.

KASTELEYN, P. W.

1961. The statistics of dimers on a lattice, I. The number of dimer arrangements on a quadratic lattice, *Physica* **27**, 1209.

1963. Dimer statistics and phase transitions, *J. Math. Phys.* **4**, 287.

KAUFMAN, B.

1949a. Crystal statistics, II. Partition function evaluated by spinor analysis, *Phys. Rev.* **76**, 1232.

1949b. and ONSAGER, L., Crystal statistics, III. Short-range order in a binary Ising lattice, *Phys. Rev.* **76**, 1244.

KOSHLAND, D. E.

1966. NEMETHY, G., and FILMER, D., Comparison of experimental binding data and theoretical models in proteins containing subunits, *Biochemistry* **5**, 365.

KRAMERS, H. A.

1941. and WANNIER, G. H., Statistics of the two-dimensional ferromagnet, I and II, *Phys. Rev.* **60**, 252, 263.

KRÖNIG, A. K.

1856. Grundzüge einer Theorie der Gase, *Ann. Phys.* **99**, 315.

LEBOWITZ, J. L.

1966. and PENROSE, O., Rigorous treatment of the van der Waals–Maxwell theory of the liquid–vapor transition, *J. Math. Phys.* **7**, 98.

1969. and LIEB, E. H., Existence of thermodynamics for real matter with Coulomb forces, *Phys. Rev. Letters* **22**, 631.

LEE, T. D.

1952. and YANG, C. N., Statistical theory of equations of state and phase transitions, II. Lattice gas and Ising model, *Phys. Rev.* **87**, 410.

LENARD, A.

1968. and DYSON, F. J., Stability of matter, II, *J. Math. Phys.* **9**, 698.

LIEB, E. H.

1966. Quantum-mechanical extension of the Lebowitz–Penrose theorem on the van der Waals theory, *J. Math. Phys.* **7**, 1016.

LOSCHMIDT, J.

1876. Über das Warmegleichgewicht eines Systems von Körpern mit Rücksicht auf die Schwere, *Wien. Ber.* **73**, 139.

LUTTINGER, J. M.

1958. and KOHN, W., Quantum theory of electrical transport phenomena, II, *Phys. Rev.* **109**, 1892.

MAXWELL, J. C.

1860. Illustrations of the dynamical theory of gases, *Phil. Mag.* Ser. 4, **19**, 19; **20**, 21, 33. [Reprinted in part in Brush (B1965), Vol. 1, p. 148.]

1867. On the dynamical theory of gases, *Phil. Trans. Roy. Soc.* (*London*) **157**, 49. [Reprinted in Brush (B1965), Vol. 2.)]

1874. On the dynamical evidence of the molecular constitution of bodies, *Nature* **11**, 53. [Also, *Collected Works*, Vol. 2, Dover, New York, p. 424.)

MONOD, J.

1965. WYMAN, J., and CHANGEUX, J. P., On the nature of allosteric transitions: A plausible model, *J. Mol. Biol.* **12**, 88.

MONTROLL, E. W.

1963. POTTS, R. B., and WARD, J. C., Correlations and spontaneous magnetization of the two-dimensional Ising model, *J. Math. Phys.* **4**, 308.

1966. and GOEL, N. S., Denaturation and renaturation of DNA, I. Equilibrium statistics of copolymeric DNA, *Biopolymers* **4**, 844.

1968. *Lectures on the Ising Model of Phase Transitions* (Brandeis Lectures 1966). Gordon & Breach, New York.

ONSAGER, L.

1944. Crystal statistics, I. A two-dimensional model with an order–disorder transition, *Phys. Rev.* **65**, 117.

1949. Discussion remark (Spontaneous magnetization of the two-dimensional Ising model) *Nuovo Cimento* (*Suppl.*) **6**, 261.

PAULING, L.

1935. The oxygen equilibrium of hemoglobin and its structural interpretation, *Proc. Natl. Acad. Sci. U.S.* **21**, 186.

PEIERLS, R.

1936a. Ising's model of ferromagnetism, *Proc. Cambridge Phil. Soc.* **32**, 477.

1936b. Statistical theory of superlattices with unequal concentrations of the components, *Proc. Roy. Soc.* (*London*) **A154**, 207.

PERUTZ, M. F.

1960. ROSSMANN, M. G., CULLIS, A. F., MUIRHEAD, H., WILL, G., and NORTH, A. C. T., Structure of haemoglobin, *Nature* **185**, 416.

1964. The hemoglobin molecule, *Sci. Am.* **211**, No. 5, 39.

RUELLE, D.

1963. Classical statistical mechanics of a system of particles, *Helv. Phys. Acta* **36**, 183.

RUMFORD (THOMPSON, B.)

1798. An inquiry concerning the source of the heat which is excited by friction, *Phil. Trans. Roy. Soc.* (*London*) **80**.

RUSHBROOKE, G. S.

1963. On the thermodynamics of the critical region for the Ising problem, *J. Chem. Phys.* **39**, 842.

1965. On the Griffiths inequality at a critical point, *J. Chem. Phys.* **43**, 3439.

SCHULTZ, T. D.

1964. MATTIS, D. C., and LIEB, E. H., Two-dimensional Ising model as a soluble problem of many fermions, *Rev. Mod. Phys.* **36**, 856.

SHERMAN, S.

1960. Combinatorial aspects of the Ising model for ferromagnetism, I. A conjecture of Feynman on paths and graphs, *J. Math. Phys.* **1**, 202.

SINAI, J.

1963. On the proof of the ergodic hypothesis for a dynamical system of statistical mechanics, *Sov. Mat. Dokl.* **4**, 1818.

1966. Ergodicity of Boltzmann's gas model in statistical mechanics foundations and applications, J.U.P.A.P. Meeting, Copenhagen, 1966, pp. 559–573.

TAKAHASHI, H.

1942. Eine einfache Methode zur behandlung der statistischen Mechanik eindimensionaler Substanzen, *Proc. Phys.-Math. Soc. (Japan)* **24**, 60. [English translation: A simple method for treating the statistical mechanics of one dimensional substances, in Lieb and Mattis (B1966), p. 25.]

TEMPERLEY, H. N. V.

1961. and FISHER, M. E., Dimer problem in statistical mechanics—an exact result, *Phil. Mag.* **6**, 1061.

THOMPSON, C. J.

1965. Algebraic derivation of the partition function of a two-dimensional Ising model, *J. Math. Phys.* **6**, 1392.

1968a. Spherical model as an instance of eigenvalue degeneracy, *J. Math. Phys.* **9**, 1059.

1968b. Models for hemoglobin and allosteric enzymes, *Biopolymers* **6**, 1101.

TONKS, L.

1936. The complete equation of state of one, two, and three dimensional gases of hard elastic spheres, *Phys. Rev.* **50**, 955.

UHLENBECK, G. E.

1963. HEMMER, P. C., and KAC, M., On the van der Waals theory of the vapor-liquid equilibrium, II. Discussion of the distribution functions, *J. Math. Phys.* **4**, 229.

VAN DER WAERDEN, B. L.

1941. Die lange Reichweite der regelmässigen Atomanordnung in Mischkristallen, *Z. Physik* **118**, 473.

VAN HOVE, L.

1949. Quelques propriétés générales de l'intégrale de configuration d'un système de particules avec interaction, *Physica* **15**, 951.

1950. Sur l'intégrale de configuration pour les systémes de particules à une dimension, *Physica*, **16**, 137.

VAN KAMPEN, N. G.

1964. Condensation of a classical gas with long range attraction, *Phys. Rev.* **135** A362.

WAKEFIELD, A. J.

1951. Statistics of the simple cubic lattice, I and II, *Cambridge Phil. Soc.* **47**, 419, 799.

WANNIER, G. H.

1945. The statistical problem in cooperative phenomena, *Rev. Mod. Phys.* **17**, 50.

WATERSTON, J. J.

1845. On the physics of media that are composed of free and elastic molecules in a state of motion. Submitted 1845; published courtesy of Lord Rayleigh in *Phil. Trans. Roy. Soc. (London)* **183A**, 5 (1893).

WEBER, K.

1968. New structural model of *E. coli* aspartate transcarbamylase and the amino-acid sequence of the regulatory polypeptide chain, *Nature* **218**, 1116.

WU, T. T.

1966. Theory of Toeplitz determinants and the spin correlations of the two dimensional Ising model, I, *Phys. Rev.* **149**, 380.

WU, T. T.

1969. Secondary structures of DNA, *Proc. Natl. Acad. Sci. U.S.* **63**, 400.

YANG, C. N.

1952a. The spontaneous magnetization of a two-dimensional Ising model, *Phys. Rev.* **85**, 809.

1952b. and LEE, T. D., Statistical theory of equations of state and phase transitions, I. Theory of condensation, *Phys. Rev.* **87**, 404.

ZERMELO, E.

1896. Über einen Satz der dynamik und der mechanischen Wärmetheorie, *Wien. Ann.* **57**, 485.

ZIMM, B. H.

1959. and BRAGG, J. K., Theory of the phase transition between helix and random coil in polypeptide chains, *J. Chem. Phys.* **31**, 526.

Reference Books

BRUSH, S. G.

1965. *Kinetic Theory*, 3 vols., Pergamon Press, Elmsford, N.Y.

BUCHDAHL, H. A.

1966. *The Concepts of Classical Thermodynamics*, Cambridge University Press, New York.

CALLEN, H. B.

1960. *Thermodynamics*, Wiley, New York.

CARLEMAN, T.

1957. *Problèmes mathématiques dans la théorie cinétique des gaz*, Almqvist and Wiksells Boktrycheri AB, Uppsala.

CERCIGNANI, C.

1969. *Mathematical Methods in Kinetic Theory*, Plenum Press, New York.

CHAPMAN, S.

1953. and COWLING, T. G., *The Mathematical Theory of Non-uniform Gases*, Cambridge University Press, New York.

EHRENFEST, P.

1911. and EHRENFEST, T., Begriffliche Grundlagen der statistischen Auffassung in der Mechanik, *Encyklopädie der mathematischen Wissenschaften*, Teubner, Bd 4, Teil 32, Liepzig. (English translation: *The Conceptual Foundations of the Statistical Approach in Mechanics*, Cornell University Press, Ithaca, N.Y.)

GANTMACHER, F. R.

1964. *The Theory of Matrices*, Vol. 2, Chelsea, New York.

GREEN, H. S.

1964. and HURST, C. A., *Order–Disorder Phenomena*, Wiley-Interscience, New York.

HARDY, G. H.

1964. LITTLEWOOD, J. E., and PÓLYA, G., *Inequalities*, Cambridge University Press, New York.

HUANG, K.

1963. *Statistical Mechanics*, Wiley, New York.

KAC, M.

1959. *Probability and Related Topics in Physical Sciences*, Wiley-Interscience, New York.

KHINCHIN, A. I.

1949. *Mathematical Foundations of Statistical Mechanics*, Dover, New York.

LIEB, E. H.

1966. and MATTIS, D. C., *Mathematical Physics in One Dimension*, Academic Press, New York.

MURNAGHAN, F. D.

1938. *The Theory of Group Representations*, Johns Hopkins Press, Baltimore.

NEWMAN, J. R.

1956. *The World of Mathematics*, Vol. II, Simon and Schuster, New York.

PIPPARD, A. B.

1957. *Elements of Classical Thermodynamics*, Cambridge University Press, New York.

ROLLER, D.

1950. *The Early Development of the Concept of Temperature and Heat*, Harvard University Press, Cambridge, Mass.

RUELLE, D.

1969. *Statistical Mechanics: Rigorous Results*, W. A. Benjamin, New York.

STANLEY, H. E.

1971. *Introduction to Phase Transitions and Critical Phenomena*, Oxford University Press, New York.

THOMSON, W.

1879. and TAIT, P. G., *Treatise on Natural Philosophy*, Cambridge University Press, New York.

UHLENBECK, G. E.

1963. and FORD, G. W., *Lectures in Statistical Mechanics*, American Mathematical Society, Providence, R.I.

WANNIER, G. H.

1966. *Statistical Physics*, Wiley, New York.

WATSON, J. D.

1968. *The Double Helix*, Atheneum, New York.

Index

Milton Keynes UK
Ingram Content Group UK Ltd.
UKHW020219300324
440331UK00006B/578

9 780691 608686